Seafood Proteins

CONTENTS

PREFACE

Fish and marine invertebrates are important sources of nutrients for the world's population, and many species have exceptionally high market value because of their exquisite sensory properties. Both the utilization of the available catch in different forms and the market price are affected by the quality of the fish. Proteins and nonprotein nitrogenous compounds play a crucial role in the nutritional value and sensory quality of seafoods as well as in the suitability of different species to various forms of processing, preservation, and use in other branches of the food industry. This role of proteins results from their basic chemical and biochemical properties and functions in different tissues. A presentation of the actual state of knowledge on seafood nitrogenous compounds in one volume may contribute to a better understanding of the involvement of these components in all stages of handling and processing fish.

It has been possible to prepare this text thanks to the cooperative effort of an international group of specialists. The editors of the book are greatly indebted to all colleagues who have willingly contributed to this volume, sharing their knowledge and experience, as well as to all persons who have granted permission to use their previously published materials.

A large part of the book has been prepared during my sabbatical in the Department of Marine Food Science, National Taiwan Ocean University (NTOU) in Keelung, Taiwan. Thanks to the generous invitation of the National Science Council of Taiwan, I had a chance to work with Professor Bonnie Sun Pan and her eager students in the competitive atmosphere of this rapidly developing university. I would like to acknowledge the friendly cooperation of all my Chinese friends in NTOU.

Last but not least, I thank my wife Krystyna for her continued understanding of my long-lasting and time-consuming hobby—writing food science books.

Zdzisław E. Sikorski

CONTRIBUTORS

Javier A. Borderias, Ph.D.
Instituto del Frio
Ciudad Universitaria
Madrid, Spain E-28040

Norman F. Haard, Ph.D.
Professor
Department of Food Science
and Technology
University of California,
Davis
Davis, CA 95616

Anna Kołakowska, Ph.D.,
D.Sc.
Professor and Head
Department of Seafood
Quality
University of Agriculture
Szczecin, Poland 71-550

Tyre C. Lanier, Ph.D.
Professor
Department of Food Science
North Carolina State University
Raleigh, NC 27695-7624

Adriaan Ruiter, Ph.D.
Professor
Department of the Science of
Food of Animal Origin
Utrecht University
Utrecht, The Netherlands

Fereidoon Shahidi, Ph.D.
Professor
Departments of Biochemistry and
Chemistry
Memorial University of
Newfoundland
St. John's, Newfoundland,
Canada A1B3X9

First published in 1994 by
Chapman & Hall
One Penn Plaza
New York, NY 10119

Published in Great Britain by
Chapman & Hall
2–6 Boundary Row
London SE1 8HN

Library of Congress Cataloging-in-Publication Data

Seafood proteins / Zdzisław E. Sikorski, Bonnie Sun Pan, Fereidoon
    Shahidi.
        p.   cm.
    Includes index.
    ISBN 0-412-98481-4
    1. Seafood—Composition.   2. Fishery processing—Quality control.
3. Muscle proteins.   I. Sikorski, Zdzisław E.   II. Pan, Bonnie Sun.
III. Shahidi, Fereidoon, 1951–   .
TX556.5.S43   1994
664'.94—dc20   93-48001
                                                                     CIP

British Library Cataloguing in Publication Data available

Please send your order for this or any other Chapman & Hall book to **Chapman & Hall, 29 West 35th Street, New York, NY 10001, Attn: Customer Service Department.** You may also call our Order Department at 1-212-244-3336
or fax your purchase order to 1-800-248-4724.

For a complete listing of Chapman & Hall's titles, send your request to **Chapman & Hall, Dept. BC, One Penn Plaza, New York, NY 10119.**

Seafood Proteins

Edited by
Zdzisław E. Sikorski, Ph.D., D.Sc.
Bonnie Sun Pan, Ph.D.
Fereidoon Shahidi, Ph.D.

CHAPMAN & HALL
New York • London

Shi-Yen Shiau, Ph.D.
Professor and Head
Department of Marine Food
  Science
National Taiwan Ocean
  University
Keelung, Taiwan 202, Republic
  of China

Zdzisław E. Sikorski, Ph.D.,
  D.Sc.
Professor and Head
Department of Food Preservation
Technical University Politechnika
  Gdańska
Gdańsk, Poland 80-952

Benjamin K. Simpson, Ph.D.
Associate Professor
Department of Food Science and
  Agriculture Chemistry
McGill University, MacDonald
  Campus
Ste-Anne-de-Bellevue
Quebec, Canada H9X1CO

Bonnie Sun Pan, Ph.D.
Professor and Dean
College of Fisheries Science
National Taiwan Ocean
  University
Keelung, Taiwan 202, Republic
  of China

1

INTRODUCTION

Zdzisław E. Sikorski and Bonnie Sun Pan

THE AVAILABILITY AND UTILIZATION OF SEAFOOD PROTEINS

The annual world catch of fish and invertebrates is actually about 100×10^6 tons and is expected to increase in the next decade by about 10%. The total marine resources available for commercial fisheries by contemporary technology are estimated to be about 200×10^6 tons. An annual catch of such size if taken in conditions respective of all conservative measures provided by fishery regulations would still not cause any significant threat of jeopardizing the biological balance in the fishing areas. The annual world harvest of fish and invertebrates from aquaculture, including mariculture, is about 10×10^6 tons and has been increased in the recent decade by over 60%. The gross contents of crude protein in the total catch is about 13×10^6 tons.

The utilization of the catch is far from optimal, as only about 70% of the total landings are intended for human consumption in different forms. If the processing offals are also taken into account, as much as about 65% of the total crude protein in the catch is used for feeding animals and even wasted. The total crude protein contained in the effluents of some fish meal plants may be as high as 30% of the N×6.25 content in the raw material. If drained into surface waters without treatment, such effluents from large plants create very serious pollution problems.

The utilization of different species of fish and invertebrates for various purposes is affected by their availability and the processing capacity of the fishing industry, as well as by the size, anatomical structure, and sensory properties of the muscles and other valuable parts of the animals. Many species of fish, crustaceans, and esoteric molluscs, for example, the sturgeon, salmon, grouper, cuttlefish, crab, lobster, and abalone, are very highly valued because of their superior eating quality. They are usually selected for the most exquisite cuisines. On the other hand, many other species because of their small size, large number of bones, dark color, or undesirable flavor and texture, for example, capelin, menhaden, or Antarctic krill, are rarely used for food in unprocessed form or are utilized for producing fish meal and oil.

The sensory quality of seafoods depends mainly on the structure and the chemical composition of the muscle tissues. These are characteristic of different species and generally undergo seasonal changes. The sensory properties and nutritive value are, to a large extent, affected by the lipids. The nitrogenous compounds, however, play a most significant role because they are responsible for the rheological characteristics, color, flavor, and biological value of the products.

THE EFFECT OF ENVIRONMENTAL FACTORS ON THE PROTEINS OF AQUATIC ORGANISMS

The aquatic organisms are exposed to environmental factors different from those that affect the land animals, mainly temperature, pressure, salt concentration, availability of oxygen, and the presence and concentration of pollutants. Therefore, many of the proteins of seafood animals have unique features which either have an impact on the properties of the muscles as food or can be utilized in other applications. The low temperature of the habitat of many species has resulted in a special adaptation of the fish to the synthesis of antifreeze glycoproteins and of their blood to the decreasing concentration of hemoglobin. The activity of proteolytic enzymes of several species of fish, molluscs, and crustaceans is seasonally extremely high due to the need of high cellular turnover in periods of intensive synthesis of genetic material. Many fish proteinases have very high activity at low temperatures and some are more active than their mammalian counterparts in a wide range of temperatures.

Characteristic of various species of sea fish and shellfish is a high activity of urease and the presence of trimethylamine demethylase, thiaminase, chitinase, and phenolase. Substantial amounts of low-molecular-weight calcium-binding proteins may be present in the muscles of fish. Molluscs contain, in their "catch muscles," a characteristic myofibrillar protein, paramyosin. Seafood collagens differ from the collagens of slaughter animals in the large amount of saccharide moieties they contain, in the structural changes due to maturation, and in thermal stability. Some low-molecular-weight nitrogenous compounds, for example, trimethylamine oxide (TMAO) and urea, have different special roles in aquatic animals.

THE DETERIORATION IN QUALITY AFTER CATCH

The quality of the catch generally undergoes rapid changes, even within hours, especially with respect to the aroma and flavor characteristics of very fresh seafoods. In unfrozen fish, the quality deterioration proceeds mainly due to postmortem biochemical processes in proteins and nonprotein nitrogenous compound (NPN), whereas, in frozen seafoods, the sensory changes are caused primarily by lipid oxidation. However, in frozen fish, proteins participate also in various processes which result in toughening and undesirable discolorations of the products (Table 1–1).

The proteins are involved in these deteriorative changes mainly as enzymes, catalyzing the catabolism of different meat constituents that are primarily responsible for the rapid loss of the high quality of very fresh seafoods. However, proteins also constitute the material of the tissues undergoing hydrolysis, polymerization, deamination, decarboxylation, oxidation, and other biochemical reactions, leading to the loss in the desirable texture of the fish and of the integrity of the fish fillets, the development of toughness and off-colors of the meat, as well as to the generation of off-flavors. The rate of quality deterioration is affected both by the species characteristics, that is, the concentration and properties of different proteins and NPN in various seafoods, as well as by the parameters of storage and processing, mainly temperature and the access of oxygen.

THE ROLE OF PROTEINS IN THE PROCESSING OF FISH

Proteins have a major effect on the suitability of various fish species from different areas and in different seasons to be processed to particular traditional or modern products. A good example is the centuries old

Table 1–1 Involvement of proteins in changes decreasing the sensory quality of seafoods

Process	Undesirable Quality Changes
Handling on board	Loss in color and desirable flavor
Refrigerated storage	Deterioration of texture, autolysis, development of off-odors, off-flavors, and off-colors
Freezing	Changes in tissue structure
Frozen storage	Oxidation and discolorations, development of off-odors and off-flavors, toughening, loss in functional properties
Manufacturing of gelled products	"Modori"-type softening of the gels
Drying	Oxidation and discolorations, development of off-odors and off-flavors, toughening, loss in functional properties
Smoking	Loss of appearance due to skin gaping
Salting	Toughening, loss in hydration, loss in texture, development of off-odors and off-flavors, development of whitish surface patches
Marinating	Loss in texture, development of off-odors and off-flavors
Canning	Honeycombing, discolorations, skin gaping, loss in hydration and texture, development of off-odors and off-flavors

experience of the European fish processors in salt-curing of different stocks of herring. The factors responsible for the exquisite quality of the lightly salted maatjes herring are presently known to be the delicate texture of the immature fish, the very active proteinases, high fat, and not too high salt content.

In modern methods of fish processing, a very significant role is played by the functional properties of fish proteins, mainly the water-holding capacity and the gelling ability. The proteins of various species of marine fish and shellfish differ significantly in this respect. However, the results of recent investigations have made it possible to control these properties by applying suitable processing parameters, as well as by chemical or biochemical modifications of the proteins.

PROTEINS AND THE NUTRITIVE VALUE OF SEAFOODS

In many countries, seafoods constitute the major source of animal protein in the diet of the population. In several developing countries, a variety of dried, fermented, and salted fish products supplies the major part of the essential amino acids required for increasing the nutritional value of the mostly carbohydrate staple foods. Seafood proteins are very well suited for this purpose. They have, as total muscle proteins, a favorable amino acid composition, resulting from a low content of collagen in the meat of the majority of species of fish and marine invertebrates. Furthermore, the edible parts of seafoods contain only marginal amounts of proteinase inhibitors. However, in very many seafood organisms, various indigenous allergenic or toxic nitrogen compounds can be found. The occurrence of these compounds is not only a species characteristic but is also affected by environmental and seasonal factors. Both aspects have been extensively investigated during the recent two decades.

EXPANDING APPLICATIONS OF SEAFOOD PROTEINS

The use of seafood proteins has been expanding during the recent two decades both in food products and in other areas of application. Due to the depletion of many valuable resources traditionally used for producing local or worldwide popular fish products, proteins from less valuable but abundant species, in form of bland minces known as surimi, are used for manufacturing traditional gelled Japanese products, as well as to make analogues of various expensive seafood items. This bland material has also been found increasingly valuable as a nutritional and functional component in other foods, including products containing meats.

During the recent two decades, extensive research has also been carried out on different industrial applications of various constituents of marine organisms. One of the most spectacular examples is the effort to find commercially viable methods of producing shellfish chitin preparations, suitable for effluent treatments such as wound-healing aids and as enzyme supports, as well as to utilize chitin as a substrate in chemical synthesis. There are also attempts to make use, in new applications, of the characteristic properties of many proteins of aquatic organisms, in particular the fish, squid, and krill enzymes, to utilize the less valuable fish species as raw material for producing peptones for microbiological purposes, and to produce gelatins and glues from marine collagen sources.

2

THE CONTENTS OF PROTEINS AND OTHER NITROGENOUS COMPOUNDS IN MARINE ANIMALS

Zdzisław E. Sikorski

THE CONTENTS OF NITROGENOUS COMPOUNDS

Proteins are very important components of fish meat because they affect both the nutritional value and the sensory properties of fish products. In the muscles and various organs of marine animals, they perform their functions, being associated with or interacting with a host of other components, mainly water, ions, lipids, and carbohydrates.

In various sources on food composition data, the contents of proteins in foods is usually given as crude protein, generally as 6.25N. This includes not only proteins but also NPN compounds. The true nitrogen-to-protein conversion factor is, however, different for the proteins in various foods because it depends on the amino acid composition and on the contents of various components of the NPN fraction. In some foods, the NPN may constitute up to 25% of total N. In many of these compounds, the C : N ratio is similar to the average in amino acids. In some of them, however, the N content may be very high, for example, in urea as high as 47%. The average conversion factor estimated recently by Sosulski and Imafidon (1990) for 23 various food products, with no correction for non-amino-acid N in the total N, is 5.68. For different classes of foods, it is in the range of 5.14–6.61. For meat and fish, the values are 5.72–5.82.

The crude protein content in different seafoods depends on the species and variety, the state of nutrition, and the stage of the reproductive cycle of the animals, as well as on the specific properties of the different parts of the organisms (Table 2–1). In the muscles, it is usually in the broad range 11–24% wet weight. A significant seasonal variation in the content of protein in the muscles of cod, found by Dambergs (1964), namely, a decrease during the spawning period by about 7% of the maximum value, was related to the physiological processes in the fish body during the reproductive activities. The skinned fillets of sand dab contain 13.6–17.7% crude proteins in September–March, but only 13–15.8% in the period of spawning of this species, that is, April–August. In the fillets of spent fish in May–August, analyzed by Nodzyński and Góralczyk (1977), the crude protein content was even as low as 5.4–12.9%. A much lesser effect of the spawning processes was found in the fillets of Greenland halibut, where the N×6.25 decreased from 12.4–15.6% to 12.5–12.9%. The contents of crude protein in the tail portion of the fillet is usually significantly lower than in the head section.

For commercial purposes, seafoods can be classified into four groups based on the percentage of crude protein contained in the meat: below 10%, 10–15%, 15–20%, and over 20%. The richest in protein is

Table 2–1 Crude protein in different parts of fish (percent wet weight)

Species	Viscera	Frame	Skin	Fillet
Hoki[a]	11.5–14.2	13.7–19.8	17.8–19.2	13.9–15.6
Red cod[a]	14.0–16.3	16.4–18.3	18.0–18.6	16.2–16.6
Hake[a]	8.5–15.5	15.2–17.2	15.9–21.7	15.0–17.4
Silver dory[b]	17.5–19.9	16.8–19.9	13.3–20.2	16.4–17.9
Alfonsino[a]	10.9–16.6	10.4–16.4	14.0–23.7	15.8–18.1
Monkfish[a]	7.5–15.5	15.5–17.3	19.0–21.3	15.1–18.2
Silver warehou[a]	9.3–14.0	14.4–18.6	11.9–15.9	17.0–18.5
Ling[b]	11.2–15.6	17.2–19.5	21.3–23.2	19.0–20.2
Gurnard[a]	16.5–20.1	17.3–20.5	21.4–25.4	18.8–20.5
Gemfish[b]	15.4–19.3	16.4–19.5	12.6–17.5	8.9–20.7
Rays bream[b]	15.8–19.1	13.1–19.8	18.0–22.3	20.1–21.2
Barracouta[a]	19.1–20.2	17.2–19.3	19.2–22.3	18.0–21.3
Blue moki[b]	13.0–22.0	18.3–20.1	20.4–25.3	19.9–21.6
Blue warehou[a]	10.1–13.1	15.1–19.9	15.6–17.8	20.0–21.6
Grey mullet[b]	8.5–16.0	18.8–25.3	18.2–29.4	20.0–22.9
Slender tuna[c]	19.1–23.9	17.1–21.1	14.4–29.6	17.4–24.4

Sources: Adapted from (a) Vlieg (1984a); (b) Vlieg (1984b); (c) Vlieg (1984c).

the flesh of the tunas, for example, 26.4±1.3% for the fillet of albacore tuna (Vlieg and Murray, 1988) and 27.2±0.8% for the white muscles of skipjack tuna (Vlieg, Habib, and Clement, 1983). In assaying the crude protein in the flesh of marine elasmobranchs, the urea N is usually subtracted from the total N. Fresh mature roe of fish is generally very rich in crude protein, in contrast to fish milt. The proportion of nitrogen in dry weight, however, is affected by the content of lipids (Table 2–2). A very large variability has been found in the amount of crude protein in the liver of fish of different species and condition, 3.5–23.5%, depending on the proportion of lipids. The swim bladders, skins, and scales of fish are characteristic of a fairly high proportion of crude protein, that is, about 18.5–37%, 19.5–33.5%, and 19.5–36.5%, respectively. In fish bones and fins, the proteins make up 10.5–20.5% and 12.5–21.0% (Zaitsev et al., 1969), respectively. The edible parts of marine invertebrates are generally, except for abalone, not very rich in protein (Table 2–3). The contents of protein in different parts of many species of fish and shellfish, based on an extensive literature survey, was published by Krzynowek and Murphy (1987). The crude protein in the meat and organs of sperm whale is in the range of values found in fish (Table 2–4), whereas that in seal meat is about 23% (Synowiecki, Hall, and Shahidi, 1992).

Table 2–2   Crude protein in fish roe and milt (in percent)

Species	Roe		Milt	
	Wet Weight	Dry Weight	Wet Weight	Dry Weight
Sturgeon[a]	26.0	58.4	17.5	46.7
Star sturgeon[a]	28.0	57.7	16.5	48.5
Chum salmon[a]	29.0	63.7	15.5	79.5
Pink salmon[a]	28.5	64.8	17.0	79.0
Rainbow trout[b]	19.0	75.5	—	—
Whitefish[b]	18.7	60.9	—	—
Bream[a]	26.0	75.4	12.5	45.5
Carp[a]	27.0	78.3	18.6	74.4
Pike-perch[a]	25.5	86.4	15.0	78.9
Cod[a]	20.0	82.6	13.0	78.8
Alaska pollack[c]	12.9	77.2	14.6	78.5
Herring	17.8[b]	82.4	17.7[a]	72.2
Grenadier[d]	13.4	72.4	9.9	78.0

Sources: Adapted from (a) Zaitsev et al. (1969); (b) Vuorela, Kaitaranta, and Linko (1979); (c) Lukash and Samofalov (1963); (d) Dvinin and Kuzmina (1984).

Table 2–3   Crude protein in the flesh of marine invertebrates (percent wet weight)

Common Name	Crude Protein	Common Name	Crude Protein
Abalone	17.0–23.0	Octopus	13.2–14.8
Clam	9.2–13.5	Squid	13.2–19.6
Cockle	11.8	Crab	15.0–18.4
Mussel	12.0–12.3	Lobster	18.2–19.2
Oyster	8.2–11.1	Shrimp	17.0–22.1
Scallop	14.8–17.7	Krill	12.0

Sources: Adapted from Bykowski and Kołodziejski (1983); Dąbrowski, Kołakowski, and Karnicka (1969), Krzynowek and Murphy (1987), Nettleton (1985), and Olley and Thrower (1977).

Table 2–4   Nitrogenous compounds in the meat and organs of antarctic sperm whale

Body Part	Crude Protein (% Wet Weight)	Collagen N (% Total N)	Nonprotein N
Meat, dorsal	23.4±1.9	23.9±6.6	17.1±4.1
Meat, ventral	23.6±2.5	23.5±12.2	10.2±3.5
Liver	21.6±1.4	25.8±8.9	14.1±8.3
Heart	17.1±2.2	34.2±5.9	17.1±6.7

Source: Adapted from Mrochkov et al. (1979).

PROTEINS

The muscles contain several classes of proteins which perform different functions in the animal organism. In the sarcoplasm and the extracellular fluids, there are the albumins, which are soluble in very dilute salt solutions. The content is usually about 30% of the total amount of proteins in the muscles, being generally higher in the meat of pelagic fish than in that of demersal species. In the muscles of some species, a large part of this protein fraction is the hemoproteins, which in the seal meat may be present in about 6% wet weight (Synowiecki, Hall, and Shahidi, 1992). The proteins of the myofibrils, soluble at ionic strength above 0.5, make up 40–60% of the total N×6.25. The yield of this fraction, obtained from fresh fish meat, depends on the extraction procedure and on the factors affecting the state of the proteins. The rest of the muscle proteins, insoluble in neutral salt solutions, are the proteins of the connective tissues. Their contents may be up to about 10% in elasmobranchs. In fish roe, the albumins participate in about 11% in the total N×6.25. The ovoglobulins called ichthulins, containing 0.4–1.1% sulphur and 0.2–0.6% phosphorus, make up about 75% of N×6.25, and the collagenous proteins of the ovum membrane make up about 13% of the crude protein. The proteins of fish milt belong mainly to protamines and histones, whereas those present in the swim bladder, bones, skin, fins, and scales are principally collagens.

The proportion of the amounts of different groups of proteins in seafoods is variable, changing in the course of the year by several percent. During maturation of the gonads and in periods of depletion, the muscles become richer in proteins insoluble in salt solutions.

THE NONPROTEIN NITROGENOUS COMPOUNDS

The contents of these compounds in the meat of marine animals depends on the species, the habitat, and life cycle effects, as well as on the state of freshness after catch.

In the meat of white fish, NPN makes up generally 9–15% of the total N, in clupeides 16–18%, and in some sharks up to 55%. The dark meat contains generally more NPN than the white or ordinary meat. In the muscles of molluscs and crustaceans, the NPN constitutes 20–25% of the total N, and in the tail meat of Atlantic krill about 20% (Bykowski and Kołodziejski, 1983).

About 95% of the total amount of NPN in the muscle of marine fish and shellfish is composed of free amino acids, imidazole dipeptides, TMAO and its degradation products, urea, guanidine compounds, nucleotides and the products of their postmortem changes, and betaines (Ikeda, 1979). The content of free amino acids in the body of oysters was found by Sakaguchi and Murata (1989) to undergo marked seasonal variations, being higher in winter than in summer. This could affect seasonal changes in palatability of the oysters. Among the free amino acids are also several nonprotein dicarboxylic amino acids, for example, pyrrolidine–2,5-dicarboxylic acid in the muscle of abalone (Sato, Sato, and Tsuchiya, 1981). The main peptides are carnosine, anserine, and balenine. The contents of RNA and DNA in the muscles of fish is about 100 mg/100 g and several mg/100 g, respectively, whereas in fish milt, the content of nucleic acids may reach about 50% of the dry matter. In 100 g of seal muscles, Synowiecki and Shahidi (1992) found 132–139 mg RNA and 53–61 mg of DNA, whereas in mechanically separated seal meat, the content of these acids is 182–226 and 109–110 mg, respectively.

REFERENCES

BYKOWSKI, P., and KOŁODZIEJSKI, W. 1983. "Właściwości Mięsa z Kryla odskorupionego Metodą Rolkową." *Bull. Sea Fisheries Inst. Gdynia.* 14 (5–6):53–57.

DAMBERGS, N. 1964. "Extractives of Fish Muscle. 4. Seasonal Variation of Fat, Water-Solubles, Protein, and Water in Cod (*gadus morhua* L.) Fillets." *J. Fish. Res. Bd. Canada.* 21:703–709.

DABROWSKI, T., KOŁAKOWSKI, E., and KARNICKA, B. 1969. "Chemical Composition of Shrimp Flesh (*Parapenaeus* spp.) and Its Nutritive Value." *J. Fish. Res. Bd. Canada.* 26:2969–2974.

DVININ, YU. F., and KUZMINA, V. I. 1984. "Tekhnokhimicheskaia Kharakteristika Severnogo Makrurusa." *Rybnoe Khozyaistvo.* 64 (3):66–69.

IKEDA, S. 1979. "Other Organic Components and Inorganic Components." pp. 111–12. In *Advances in Fish Science and Technology,* edited by J. J. Connell and the staff of Torry Research Station. Farnham, Surrey: Fishing News Books.

KRZYNOWEK, J., and MURPHY, J. 1987. "Proximate Composition, Energy, Fatty Acid, Sodium, and Cholesterol Content of Finfish, Shellfish, and Their Products." NOAA Technical Report NMFS 55. Washington, DC: U. S. Department of Commerce.

LUKASH, E. G., and SAMOFALOV, P. E. 1963. "Obrabotka Ikry i Molok Ryb." Pp. 377–421. In *Spravochnik Tekhnologa Rybnoi Promyshlennosti,* Vol. 1, edited by V. M. Novikov, Moskva: Pishchepromizdat.

MROCHKOV, K. A., KOVROV, G. V., PERMYAKOVA, O. N., and SHEPELEVA, G. S. 1979. "Sostav Azotistykh Veshchestv Tushi Antarkticheskogo Kashalota i ich Ratsionalnoe Ispolzovanie." *Trudy Vsesoyuznogo Nauchno–Issledovatelskogo Inst. Morskogo Rybnogo Khozyaistva Okeanografii.* 139:69–76.

NETTLETON, J. A. 1985. *Seafood Nutrition: Facts, Issues and Marketing of Nutrition in Fish and Shellfish.* New York: Van Nostrand Reinhold.

NODZYNSKI, J., and GORALCZYK, A. 1977. "Charakterystyka Biologiczna i Technologiczna Niegładzicy i Halibuta Niebieskiego z Północno-Zachodniego Atlantyku. *Studia i Materiały. Morski Inst. Rybacki. Ser. D* 13:1–69.

OLLEY, J., and THROWER, S. J. 1977. "Abalone—an Esoteric Food. Pp. 143–186. In *Advances in Food Research,* Vol. 23, edited by C. O. Chichester, E. M. Mrak, and G. F. Stewart. New York: Academic Press.

SAKAGUCHI, M., and MURATA, M. 1989. "Seasonal Variations of Free Amino Acids in Oyster Whole Body and Adductor Muscle. *Nippon Suisan Gakkaishi.* 55:2037–2041.

SATO, M., SATO, Y., and TSUCHIYA, Y. 1981. "Isolation of L-Pyrrolidine–2,5-Dicarboxylic Acid from the Muscle of Abalone." *Bull. Japan. Soc. Sci. Fish.* 47:1605–1608.

SOSULSKI, F. W., and IMAFIDON, G. I. 1990. "Amino Acid Composition and Nitrogen-to-Protein Conversion Factors for Animal and Plant Foods." *J. Agric. Food Chem.* 23:1351–1356.

SYNOWIECKI, J., and SHAHIDI, F. 1992. "Nucleic Acid Content of Seal Meat. *Food Chem.* 43:1–2.

SYNOWIECKI, J., HALL, D., and SHAHIDI, F. 1992. "Effect of Washing on the Amino Acid Composition and Connective Tissue Content of Seal Meat. *J. Muscle Foods.* 3:25–31.

VLIEG, P. 1984a. "Proximate Analysis of 10 Commercial New Zealand Fish Species." *New Zealand J. Sci.* 27:99–104.

VLIEG, P. 1984b. "Proximate Analysis of Commercial New Zealand Fish Species. 2." *New Zealand J. Sci.* 27:427–438.

VLIEG, P. 1984c. "Proximate Composition of New Zealand Slender Tuna *Allothunnus fallai. New Zealand J. Sci.* 27:435–438.

VLIEG, P., and MURRAY, T. 1988. "Proximate Composition of Albacore Tuna. *Thunnus alalunga,* from the Temperate South Pacific and Tasman Sea. *New Zealand J. Marine Freshwater Res.* 22:491–496.

VLIEG, P., HABIB, G., and CLEMENT, I. T. 1983. "Proximate Composition of Skipjack Tuna *Katsuwonus pelamis* from New Zealand and New Caledonian Waters." *New Zealand J. Sci.* 26:243–250.

VUORELA, R., KAITARANTA, J., and LINKO, R. R. 1979. "Proximate Composition of Fish Roe in Relation to Maturity." *Can. Inst. Food Sci. Technol. J.* 12:186–188.

ZAITSEV, V., KIZEVETTER, I., LAGUNOV, L., MAKAROVA, T., MINDER, L., and PODSEVALOV, V. 1969. *Fish Curing and Processing.* Moscow: MIR Publishers.

3

SARCOPLASMIC PROTEINS AND OTHER NITROGENOUS COMPOUNDS

Norman F. Haard, Benjamin K. Simpson, and Bonnie Sun Pan

INTRODUCTION

Scope

Muscles free of depot fat normally contain about 20% nitrogenous compounds. Muscle tissue contains many different kinds of nitrogenous molecules ranging in molecular weight from more than 100,000 to less than 100 daltons (Chapter 2). In number, most of these compounds are classified as sarcoplasmic components because they are soluble in the muscle sarcoplasm in contrast to the proteins in the contractile apparatus (Chapter 4) and connective tissues (Chapter 5). Chemically, nitrogen-containing compounds in fish and shellfish sarcoplasm include water and dilute salt-soluble proteins, peptides, amino acids, amines, amine oxides, guanidine compounds, quaternary ammonium compounds, purines, and urea. Sarcoplasmic proteins normally consist of about 20–25% of total fish muscle protein. The nonprotein-N content of fish muscle is normally higher than that of terrestrial animals ranging from 10–40% of the total nitrogen content.

The physiological roles of these nitrogenous compounds for the living animal are wide ranging and include functions in enzyme catalysis, osmoregulation, antifreeze, intermediary metabolism, nitrogen storage, cell structure, and transport processes. Nitrogenous compounds are important to the seafood technologist because they may directly or indirectly influence the edible qualities of fresh or processed seafood, that is, color, flavor, texture, nutrition, and safety. The purpose of this chapter is to briefly discuss our knowledge of bioactive nitrogen compounds with reference to seafood quality.

Interspecific Factors

Seafood comes from a complex taxonomic array of organisms, that is Arthropods (e.g., lobsters, crabs, shrimps), Molluscs (e.g., squids, clams, abalones), and Chordates (e.g., bony and cartilaginous fish). The major subclass of bony fish (Teleosti) consists of about 50 taxonomic orders with about 7000 freshwater and 13,000 saltwater species known at this time. Genetic diversity also exists within specie (stocks or strains) through natural genetic selection in the wild and broodstock selection of domesticated fish. Bioactive substances, notably enzymes, can differ between fish strains or stocks. Indeed, enzyme polymorphism has been used to distinguish fish strains (Zajicek, 1991).

Despite the broad taxonomic diversity of animals that comprise seafood, the basic composition of nitrogenous compounds in muscle from aquatic animals is remarkably similar. The same metabolic patterns seem to be used throughout the animal kingdom (Dixon and Webb, 1979). Differences in metabolism are normally not the result of a change in an entire metabolic pathway. Genetic or phenotypic variation in metabolism can be explained by one or more of the following: (1) occasional "all or none" addition or deletion of an enzyme to a basic pathway, (2) differences in the amount of an enzyme, (3) multiple molecular forms of an enzyme in the same tissue from the same organism, and (4) homologous forms of an enzyme in the same tissue from different organisms. Exceptional enzymes that are found only in specific sources of seafood include polyphenoloxidase in crustacean species, carnosinase in eels, and trimethylamine demethylase in gadoid fish. Similarly, a high concentration of other nitrogenous compound(s) is sometimes characteristic of muscle tissue from specific taxonomic groups, for example, histidine in the Scombroid fish and urea in Elasmobranchs.

Intraspecific Factors

Muscle is a heterogeneous tissue, normally composed of mixtures of fiber types and an extracellular matrix and containing macrophages and residual blood. The composition of nitrogenous compounds varies from muscle to muscle in the same animal. For some species, the dark muscle contains less sarcoplasmic protein (Suzuki, 1981) and more non-protein nitrogen (Haard, 1992a) than white muscle. To further compli-cate matters, the quantity, physicochemical properties, and catalytic prop-erties of enzymes (Haard, 1990) and other constituents (Haard, 1992) can vary with biological age, diet, exercise, habitat temperature, water depth, and other ontogenic and environmental factors. For example, a proteinase from winter and spring burbot differs in thermal stability, kinetic constants, and inhibitor sensitivity (Wiggs, 1974). The importance of habitat conditions is also reflected in findings that the amounts of free amino acids and other NPN differ considerably in cultured and wild fish of the same specie [reviewed by Haard (1992a)].

SARCOPLASMIC PROTEINS

Classification

The sarcoplasmic or "myogen" fraction of muscle is a large family of proteins that share the property of being soluble in water and dilute salt solution. The sarcoplasmic proteins from fish are more or less like those from land animals, that is, they include myoglobin, hundreds of enzymes, and other albumins. The muscles of fish and lower vertebrates, unlike land animals, also contain large quantities of Ca^{2+}–binding pro-teins called parvalbumins [reviewed by Gazzaz and Rasco (1992)]. The content of sarcoplasmic protein is generally higher in pelagic fish, such as sardine and mackerel, and lower in demersal fish like plaice and snapper (Table 3–1). When separated by electrophoresis or isoelectric focusing, sarcoplasmic proteins can be used to "fingerprint" or identify fish species (Toom, Ward, and Weber, 1982). Small quantities of sarco-plasmic proteins have an adverse effect on the strength and deformability of myofibril protein gels. These proteins may interfere with myosin cross-linking during gel matrix formation because they do not form gels and have poor water-holding capacity (Smith, 1991).

Enzymes

General.   Fish sarcoplasmic proteins have, for the most part, the same basic, functional classes of enzymes common in mammals, birds, and reptiles. These include the sarcoplasmic enzymes involved with phys-iological events of respiration, intracellular digestion, cell growth, cell division, and secondary metabolism. On this basis, we can surmise that

Table 3–1   Sarcoplasmic protein content in muscle of different fish

Fish	Genus Species	Sarcoplasmic Protein (g/100 g)
Mackerel	*Scomber japonicus*	6.4–8.8
Yellowtail	*Seriola quinqueradiata*	8.4–8.6
Red barracuda	*Sphyraena pinguis*	7.1
Horse mackerel	*Trachurus japonicus*	5.0–5.6
Sea bass	*Sterolepsis ischinagi*	4.2–5.7
Lizard fish	*Saurida undosquamis*	5.6
Flatfish	*Kareius bicoloratus*	4.4
Grub fish	*Neopercis sexfasciata*	4.9
Anchovy	*Engraulis japonica*	3.8

Source: Adapted from Suzuki (1981).

all six classes of enzymes are present in fish muscle. The main enzyme groups that are recognized by seafood technologists for their impact on the edible qualities of fish are hydrolases, oxidoreductases, and transferases. This subject was recently reviewed by Simpson, Smith, and Haard (1991).

Hydrolases.   Hydrolytic enzymes of importance in postharvest fish include proteinases and peptidases, lipases and phospholipases, and glycogen hydrolases. Constitutive proteinases important in seafood include those present in (1) muscle cells, (2) the extracellular matrix and connective tissue surrounding muscle cells, and (3) digestive and other organs.

The characterization of fish proteinases started with digestive proteases, initially studied more than 100 years ago (Stirling, 1884). Some digestive enzymes like salmon pepsin were purified more than 50 years ago (Norris and Elam, 1940). In more recent years, we have gained a better understanding of multiple molecular forms of digestive enzymes (Squires et al., 1986; Simpson and Haard, 1987b) and their comparative biochemistry (Simpson and Haard, 1987a; Gildberg, 1988). The presence of lysosomal enzymes in fish muscle has been known for almost 30 years (Siebert 1962; Tappel et al., 1963; Inaba et al., 1976) although new catheptic enzymes continue to be identified in fish and shellfish (Hara et al., 1988; Zeef and Dennison 1988; Ueno et al., 1988). The high-molecular-weight, heat-stable, alkaline muscle proteinases were identified in fish muscle nearly 20 years ago (Iwata et al., 1973) and have been extensively studied in connection with surimi gel weakening. More recently identified proteinases from seafood tissues include Ca^{2+}-activated neutral proteases (Jiang et al., 1990), modori-inducing neutral proteinases (Kinoshita et al., 1990), multicatalytic proteinases or proteosomes (Folco et al., 1989), and metalloproteinases with collagen-degrading activity (Bracho and Haard, 1991). The reader should be cautioned that several groups of fish proteases are not sarcoplasmic proteins but may be associated with organelles (e.g., cathepsins), connective tissue (collagenases), and myofibrils (e.g., alkaline proteinases). The literature on endogenous proteinases from seafood is reviewed in detail by Haard (1992b). Examples of fish tissues known to contain specific proteinases are summarized in Table 3–2.

Lipolytic enzymes from fish include those from digestive organs and from fish muscle (Table 3–3). Lipolysis is especially important because it contributes to deterioration in frozen fish (Haard, 1992c). Although soluble phospholipase has been isolated from pollack muscle (Audley et al., 1978), most investigators have identified muscle phospholipase with the microsomal fraction, and muscle lipase has been identified with lysosomes (Geromel and Montgomery 1980). The subject of lipolysis in fish muscle is reviewed by Shewfelt (1981).

Carbohydrate hydrolases appear to contribute to glycogen degradation in postharvest fish muscle [reviewed by Eskin (1990)]. Fish muscle appears to differ from that of terrestrial animals in its capacity to mobilize glycogen via an hydrolytic, amylaselike, pathway as well as by the well-known phosphorylase-catalyzed route of glycogen catabolism. Muscle lysosomes contain proteoglycan-degrading enzymes like β-glucuronidase and β-N-acetylhexosaminidase which may contribute to postharvest softening of muscle texture (Kim and Haard, 1992). Marine organisms also possess various digestive polysaccharide-degrading enzymes (Yamada, Takano, and Kamoi 1991; Anzai et al., 1991a, 1991b, 1992).

Oxidoreductases. Examples of oxidoreductase enzymes identified in fish tissues are summarized in Table 3–4. Polyphenol oxidases are particularly important in Crustacea because they can cause postharvest discoloration. Lipoxygenases have been identified in fish skin, gill, and muscle tissue. These enzymes have been implicated in skin carotenoid bleaching (Tsukuda, 1970) and fresh fish aroma (Josephson and Lindsey, 1986). Some fish lipoxygenase activity is associated with microsomal membranes in addition to the soluble sarcoplasmic proteins [reviewed by Haard (1992c)]. Lipoxygenase has also been identified in the hemolymph of shrimp (Kuo and Pan, 1992) and in blood platelets of fish (Chen, 1993). Lipoxygenase activity is also present in the eggs of sea urchin and starfish where it is involved with arachidonic acid metabolism and oocyte maturation (Hawkins and Brash, 1987; Meijer et al., 1986; Meijer, Maclouf, and Bryant, 1986b).

Table 3–2   Proteinases identified in fish and shellfish

Proteinase	Source	Tissue	Ref.
Cathepsins	Albacore	Muscle	1
	Carp	Muscle	2
	Cod	Muscle	3
	Salmon	Muscle	4
	Pacific sole	Muscle	5
	Rockfish	Muscle	6
	Tilapia	Muscle	7
	Squid	Hepatopancreas	8
Alkaline proteases	Filefish	Muscle	9
	White croaker	Muscle	9
	Anchovy	Muscle	9
	Rockfish	Muscle	9
	Yellowtail	Muscle	9
	Tiger shrimp	Muscle	9
	Garfish	Muscle	9
Collagenase	Rockfish	Muscle	10
	Crab	Hepatopancreas	11
	Trout	Pyloric ceca	12

References: (1) Groninger (1964); (2) Hara, Suzumatsu, and Ishihara (1988); (3) Schmitt and Siebert (1967); (4) Ting et al. (1968); (5) Geist and Crawford (1974); (6) Kim and Haard (1992); (7) Sherekar et al. (1988); (8) Hameed and Haard (1985); (9) Iwata et al. (1974); (10) Bracho and Haard (1991); (11) Eisen and Jeffrey (1969); (12) Yoshinaka et al. (1978); (13) Tanji et al. (1988); (14) Fruton and Bergmann (1940); (15) Merrett et al. (1969); (16) Kubota et al. (1970); (17) Owen and Wiggs (1971); (18) Noda

Other examples of oxidoreductases found in fish include peroxidases (Ilgner and Woods, 1985), catalase and superoxide dismutase (Scott and Harrington, 1990). Oxidoreductases are also important components of the glycolytic pathway. Although these enzymes are normally considered to be sarcoplasmic proteins, they may be particulate bound in fish muscle (Brooks and Storey, 1988a; Lushchak, 1991). Lactic dehydrogenases from marine fish are particularly well studied, although this enzyme is absent in some invertebrates like squid.

Other Enzymes.   Among the many other soluble enzymes known to occur in fish are the transferases, phosphorylase (Assaf and Graves, 1969), phosphofructokinase (Brooks and Storey, 1988b), transglutaminase (Kishi et al., 1991), and glutathione S-transferase (Nimmo and Spalding, 1985). Transglutaminase catalyzes acyl transfer reactions between γ-carboxamide groups of glutamine residues within proteins and suitable

Table 3–2 *(Cont'd.)*

Proteinase	Source	Tissue	Ref.
Pepsins	Tuna	Gastric mucosa	13
	Salmon	Gastric mucosa	14
	Dogfish	Gastric mucosa	15
	Bonita	Gastric mucosa	16
	Trout	Gastric mucosa	17
	Sardine	Gastric mucosa	18
	Hake	Gastric mucosa	19
	Cod	Gastric mucosa	20
	Smelt	Gastric mucosa	21
Trypsin	Cod	Pyloric ceca	22
	Sardine	Pyloric ceca	23
	Goldfish	Hepatopancreas	24
	Capelin	Intestine	25
	Catfish	Intestine	26
	Rainbow trout	Intestine	27
	Chinook salmon	Intestine	28
	African lungfish	Pancreas	29
Chymotrypsin	Spiny digfish	Pancreas	30
	Herring	Intestine	31
	Capelin	Intestine	31

and Murakami (1981); (19) Sanchez-Chiang and Ponce (1981); (20) Squires et al. (1986); (21) Haard et al. (1982); (22) Simpson and Haard (1984); (23) Murakami and Noda (1981); (24) Jany (1976); (25) Hjelmeland and Raa (1982); (26) Yoshinaka et al. (1984); (27) Kitamakado and Tachino (1960); (28) Croston (1965); (29) de Haen et al. (1977); (30) Ramakrishna et al. (1987); (31) Kalac (1978).

acceptors, usually primary amines. When lysine residues in protein are acyl acceptors, the reaction results in the formation of an ε-(γ-glutamyl)-lysine isopeptide bond that serves to cross-link proteins. This reaction appears to be responsible for myosin heavy-chain intermolecular cross-linking during the "suwari" process of surimi paste (Numakura et al., 1985).

TMAO demethylase, which appears to contribute to texture deterioration of frozen gadoid fish, has been isolated from various fish tissues including pyloric ceca, liver, kidney, hepatopancreas, and muscle [reviewed by Haard (1992c)]. Although there are reports of soluble TMAO demethylase from fish muscle (Tokunaga, 1965), most research has shown the enzyme to be particulate bound (Parkin and Hultin, 1981; Lundstrom et al., 1982). Some enzymes may also have a positive influence on texture of cooked fish products. For example, kidney extract from

Table 3–3   Lipolytic enzymes identified in fish and shellfish

Enzyme	Source	Tissue	Ref.
Lipase	Skate	Pancreas	1
	Trout	Pancreas	2
	Cod	Digestive tract	3
	Crayfish	Digestive tract	4
	Lobster	Gastric Juice	5
	Trout	Muscle	6
	Cod	Muscle	7
Phospholipase	Cod	Muscle	8
	Trout	Muscle	9
	Pollack	Muscle	10
	Flounder	Muscle	11

References: (1) Brockerhoff and Hoyle (1965); (2) Brockerhoff (1966); (3) Leger (1972); (4) Berner and Hammond (1970); (5) Brockerhoff (1967); (6) Geromel and Montgomery (1980); (7) Olley, et al. (1962); (8) Chawla and Ablett (1987); (9) Han and Liston (1987); (10) Audley et al. (1978); (11) Shewfelt et al. (1981).

tropical sardine (*Sardinella gibbosa*) enhances the gel strength of surimi (Siang et al., 1991).

Pigments

General.   A variety of nitrogenous compounds contribute to the color of fish and shellfish. The majority of these compounds are associated with epithelial tissue and are reviewed by Haard (1992d). The nitrogen-containing pigments include the blue-green to red carotenoproteins, blue-purple indigoids, brown-black melanins and melanoproteins, variously colored ommatids that are polycyclic, aromatic compounds from cephalopods, red-green tetrapyrroles such as biliverdin, flavins associated with the skin of scaleless fishes, purines responsible for white iridescence in leukophores of fish skin and pterins that contribute to blue iridescence in fish scales. The heme proteins are sarcoplasmic proteins that occur in the meat and influence postharvest color change and lipid oxidation.

Heme Proteins.   The pigments associated with the bright red surface color of fish meat include oxymyoglobin and oxyhemoglobin. Normally, hemoglobin contributes less to the appearance of seafood because it is lost rather easily during handling and storage, whereas myoglobin is retained by the cellular structure. Hemoglobin is absent from some crustacea and mollusks and from certain Antarctic fish with colorless

Table 3–4 Some oxidoreductase enzymes identified in fish and shellfish tissues

Enzyme	Source	Tissue	Ref.
Phenol oxidase	Lobster	Cuticle	1
	Crab	Cuticle	2
	Lobster	Shell	3
	Shrimp	Shell	4
	Shrimp	Head	4
	Mussel	Digestive tract	5
	Goldfish	Skin	6
	Octopus	Ink gland	7
	Cuttlefish	Ink gland	7
	Squid	Ink gland	7
Lipoxygenase	Various finfish	Skin	8
	Trout	Skin	9
	Herring	Muscle	10
	Trout	Muscle	11
	Flounder	Muscle	12
9-Lipoxygenase	Trout	Gill	13
12-Lipoxygenase	Trout	Gill	14
15-Lipoxygenase	Trout	Gill	15
Peroxidase	Crayfish	Hepatopanreas	16
Superoxide dismutase	Salmon	Erythrocytes	17
Catalase	Salmon	Erythrocytes	17
Oactate dehydrogenase	Tuna	Muscle	18
	Halibut	Muscle	18
	Shrimp	Muscle	19
	Trout	Muscle	20
	Skate	Muscle	21
D-Glyceraldehyde-3-phosph. dehydrogenase	Lobster	Muscle	22
	Cod	Muscle	22
Glutamate dehydrogenase	Muscle	Hepatopancreas	23
	Mollusks	Muscle	24
Uricase	Various finfish	Liver	25

References: (1) Ferrer et al. (1989); (2) Summers (1967); (3) Savagaon and Sreenivasan (1975); (4) Simpson et al. (1988); (5) Waite (1985); (6) Kim and Tchen (1962); (7) Prota et al. (1981); (8) Tsukuda (1970); (9) Hsieh and Kinsella (1986); (10) Slabyi and Hultin (1982); (11) Han and Liston (1987); (12) Macdonald and Hultin (1987); (13) Winkler et al. (1992); (14) German and Berger (1990); (15) German and Crevling (1990); (16) Ilgner and Woods (1985); (17) Scott and Harrington (1990); (18) Low et al. (1973); (19) Thebault (1981); (20) Henry and Ferguson (1985); (21) Luschak (1991); (22) Cowey (1967); (23) Ruano et al. (1985); (24) Wickes and Morgan (1976); (25) Kinsella et al. (1985).

blood (Love, 1980). The prosthetic group in hemoglobins, protoheme 1, is the same in all animals; however, the protein part of the molecule more or less differs from specie to specie. Most fish hemoglobins are tetrameric proteins; however, invertebrate hemoglobins often contain more than four heme groups and primitive fishes (lampreys and hagfish) have monomeric, myoglobin-like hemoglobin. The combination of swordfish hemoglobin with hydrogen sulfide can cause green discoloration and off-odor due to isovalerianic acid formation (Amano, 1949). Hemoglobin and myoglobin of tuna tend to become insoluble during frozen storage, even at −80°C (Chow, Ochiai, and Hashimoto, 1985; Chow et al., 1987). Formation of myoglobin dimer appears to occur via a disulfide bond, and insolubilization is facilitated by myoglobin decomposition.

Myoglobin.   Myoglobin is a globular heme protein with a molecular weight of about 18 kDa. Myoglobins from fish have an amino acid composition that differs considerably from myoglobin of mammals (Brown, 1962). A salient feature of some fish myoglobins (e.g., tuna) is that they contain a cysteine residue adjacent to the heme. Such fish myoglobins can react with thiol compounds in the presence of TMAO to cause green discoloration of the meat (Yamagata et al., 1971). In general, fish myoglobins are more readily oxidized to metmyoglobin than mammalian myoglobins [reviewed by Haard (1992d)].

Hemocyanins.   Shellfish contain copper-containing proteins called hemocyanins that, like hemoglobins, combine reversibly with oxygen. Hemocyanin binds one molecule of oxygen to two copper atoms and has a functional subunit of 50–75 kDa. Heat-denatured hemocyanin or oxyhemocyanin turns white and in the presence of hydrogen sulfide forms a blue-green color. Hemocyanin also possess catalaselike activity (Ghiretti, 1956).

Parvalbumins

Parvalbumins are a group of heat-stable, water-soluble, acidic, low-molecular-weight (about 12 kDa) Ca^{+2}-binding proteins. Parvalbumin and another calcium-binding protein calmodulin occur in the muscle of aquatic lower vertebrates, including fish. In fish, these molecules may constitute 20–30% of the sarcoplasmic myogen fraction (Schwimmer, 1981). The calcium-binding proteins appear to serve the same function as the troponin complex and tropomyosin in higher vertebrates (Grandjean et al., 1977). However, the exact physiological role of parvalbumins is not clear at this time. They are associated with fast, nerve-impulse-activated skeletal muscle (Pechere, 1977). Reports of parvalbumin isolation from fish muscle are summarized in Table 3–5.

Interestingly, the primary structures of fish parvalbumins are highly homologous with cod allergen M and is now classified as a parvalbumin (Elsayed and Bennich, 1975; Fig. 3–1). Allergen M is one of the most potent food allergens (Aas, 1987). The immunology of fish parvalbumins was reviewed recently by Gazzaz and Rasco (1992). The immunologically active sites of allergen M are (1) residues 33–34, adjacent to the first Ca^{2+}-binding site, (2) residues 65–74, on the junction between the two Ca^{2+}-binding sites, and (3) residues 88–96, on the second Ca^{2+} bind loop of the molecule. Crustacean IgE-mediated hypersensitivities are similar to those of finfish; however, the allergens (Antigen I and II in shrimp) are different (Lehrer and McCants, 1987; Lehrer and Daul, 1992). Shrimp antigen I is a heat-stable protein of 20.5 kDa containing 0.5% carbohydrate. Antigen II is 38.3 kDa with 4% carbohydrate and stable in shrimp boiled for 10 min.

The parvalbumins are also of interest to seafood technologists because they can be used to confirm species of origin of processed fish

Table 3–5 Parvalbumins isolated from the skeletal muscle of fish

Source	Components	Ref.
Carp	3	1
Hake	1 major	2
Pike	1 major	3
Tilapia	2	4
Lungfish	5	5
Raja sp.	1 major	6
Perch	2 major	7
Cod	2	8
Oryzias sp.	5	9
Lepomis sp.	5	10
Cyprinidea family	5	11
Toadfish	1	12
Eel	3	13
Salmonidae family	5	14
Pollack	5	15

References: (1) Konosu et al. (1965); (2) Pechere et al. Capony (1971); (3) Frankenne et al. (1973); (4) Piront et al. (1968); (5) Elayed and Bennich (1975); (6) Gerday and Teuwis (1972); (7) Lehky and Stein (1979); (8) Bhushana et al. (1969); (9) Sakaizumi (1985); (10) Whitmore (1986); (11) Bobak and Slechta (1988); (12) Gerday et al. (1989); (13) Dubois and Gerday (1990); (14) Rehbein and van Lessen (1989); (15) Woychik and Dower (1990).

Figure 3–1 Amino acid sequences of cod allergen M and other parvalbumins.

Residue	1	2	3	4	5	6	7	8	9	10	11	12	13	14	15	16	17	18	19	20	21	22	23	24	25	26	27	28	29	30
Allergen M	A	F	A	K	G	I	L	S	N	A	D	I	K	A	A	E	A	F	K	E	G	S	F	D	E	D	G	F	Y	F
Whiting	A	F	A	K	G	I	L	A	F	D	I	K	A	A	V	K	A	C	V	A	E	G	S	F	S	Y	D	G	F	F
Hake	S	F	K	I	L	L	K	S	K	D	I	K	A	A	L	K	A	C	Z	A	B	S	D	H	Y	K	G	E	F	F
Pike II	S	F	A	G	L	K	D	A	A	I	T	A	A	L	A	L	C	A	V	A	E	G	D	F	K	H	K	A	F	F
Pike III	A	F	A	G	V	L	N	D	A	D	I	A	A	A	L	E	A	C	K	A	A	D	S	F	N	H	K	A	F	F
Carp III	A	F	A	G	V	L	N	D	A	D	I	A	A	A	L	E	A	C	K	A	A	D	S	F	N	H	K	A	F	F
CarpIV	A	F	A	G	V	L	N	D	A	D	I	T	A	A	L	E	A	C	K	A	A	D	S	F	N	H	K	A	F	F

Residue	31	32	33	34	35	36	37	38	39	40	41	42	43	44	45	46	47	48	49	50	51	52	53	54	55	56	57	58	59	60
Allergen M	A	K	V	G	L	D	A	F	L	A	F	D	L	K	K	L	F	C	K	A	D	E	D	K	E	G	F	D	E	E
Whiting	A	K	C	S	L	S	G	K	S	A	B	B	B	V	K	V	V	C	I	I	D	Z	D	K	S	G	F	L	V	E
Hake	T	K	I	G	L	K	S	K	S	A	A	D	I	K	K	L	F	G	I	I	D	D	D	K	S	D	F	I	I	E
Pike II	A	K	V	G	L	K	A	M	S	A	A	D	V	K	K	V	F	Y	I	I	D	Q	D	A	S	G	F	I	E	E
Pike III	A	K	V	G	L	T	S	K	S	A	D	D	V	K	K	A	F	A	I	I	D	Q	D	K	S	G	F	I	E	E
Carp III	A	K	V	G	L	T	S	K	S	A	D	D	V	K	K	A	F	A	I	I	D	Q	D	K	S	G	F	I	E	E
CarpIV	A	K	V	G	L	T	S	K	S	A	D	D	V	K	K	A	F	A	I	I	D	Q	D	K	S	G	F	I	E	E

Residue	61	62	63	64	65	66	67	68	69	70	71	72	73	74	75	76	77	78	79	80	81	82	83	84	85	86	87	88	89	90
Allergen M	D	E	L	K	L	F	L	I	A	F	K	A	D	L	R	A	L	T	D	A	E	T	K	A	F	L	K	A	G	D
Whiting	B	E	L	K	L	F	L	E	V	F	K	A	G	A	R	A	L	T	D	A	E	T	K	A	F	L	K	A	G	D
Hake	N	E	L	K	L	F	L	Q	N	F	S	G	S	G	A	A	L	T	D	A	E	T	K	T	F	L	A	D	G	D
Pike II	F	E	L	K	F	V	L	Q	N	F	K	A	D	A	R	D	L	T	D	G	E	T	K	T	F	L	K	A	G	D
Pike III	D	E	L	K	L	F	L	Q	N	F	K	A	D	A	R	A	L	T	D	G	E	T	K	T	F	L	K	A	G	D
Carp III	D	E	L	K	L	F	L	Q	N	F	K	A	D	A	R	A	L	T	D	G	E	T	K	T	F	L	K	A	G	D
CarpIV	D	E	L	K	G	L	F	L	K	N	F	K	A	G	R	A	L	T	D	G	E	T	K	T	F	L	K	A	G	F

Residue	91	92	93	94	95	96	97	98	99	100	101	102	103	104	105	106	107	108	109	110	111	112	113	114	115
Allergen M	S	D	G	D	G	K	I	G	V	D	E	F	G	A	L	V	D	K	W	G	A	K	G		
Whiting	S	D	G	D	G	I	I	G	E	E	E	W	A	A	L	V	K	A	A	K					
Hake	S	D	G	D	G	M	I	G	V	F	F	E	F	A	M	V	K	G							
Pike II	K	D	G	D	G	K	I	G	D	D	E	F	T	E	T	I	H	K	A						
Pike III	K	D	G	D	G	K	I	G	I	D	E	F	A	L	L	V	H	E	A						
Carp III	S	D	G	D	G	K	I	G	G	D	E	F	T	T	L	V	K	A							
CarpIV	D	D	G	D	G	I	I	G	V	D	E	F	T	L	V	K	A								

A(alanine), B(asparagine), C(cysteine), D(aspartic), E(glutamic), F(phenylalanine), G(glycine), H(histidine), I(isoleucine), K(lysine), L(leucine), M(methionine), N(asparagine), P(proline), Q(glutamine), R(arginine), S(serine), T(threonine), V(valine), W(tryptophan), Y(tyrosine). The N terminus of all these parvalbumins is acetylated.

Source: Adapted from Elsayed and Bennich (1975); Garazz and Rasco (1992).

24

products (Woychik and Dower, 1990). It is also possible that removal of parvalbumins during surimi manufacture contributed to the improved gelation properties of water-washed fish mince.

Antifreeze Proteins

Fish from Arctic and Antarctic waters have proteins in their blood that lower freezing temperature with no significant effect on osmotic pressure and, hence, show hysteresis between melting and freezing temperatures [reviewed by Feeney (1988)]. Antifreeze glycoproteins (AFGP) are essentially polymers of $H_2N[Ala-Ala(\beta$-galactosyl(1–3)α-N-acetylga-lactosamine)Thr$]_n$Ala-Ala-COOH. AFGPs 1 to 8 have $n = 50, 45, 35,$ 28, 17, 12, and 6–4, respectively. AFGPs 6–8 constitute 80% of the total proteins and are found in several Antarctic and Arctic fish families. Other antifreeze proteins contain no carbohydrate. Some of these are very similar to the AFGPs in having two-thirds of their amino acid residues as alanine, but with other amino acids in place of carbohydrate.

In northern temperate fish like Atlantic cod, antifreeze proteins are formed in fish living at low water temperature (e.g., 0°C) and their occurrence is known to influence the freezing point of the muscle (Simpson and Haard, 1987). Cod fillets, from cold-acclimated fish, that contain antifreeze protein exhibit less quality deterioration when held under partial freezing conditions (−3°C). A patent was recently issued to use antifreeze polypeptides, or their fusion proteins, to increase the shelf life of foodstuffs (Warren et al., 1992).

NONPROTEIN-NITROGENOUS COMPOUNDS

General

The NPN content of seafood is significantly higher than that in other food myosystems, that is, 9–18% of total nitrogen in teleosts and 33–38% in elasmobranchs (Belitz and Grosch, 1987). The major classes of sarcoplasmic components are free amino acids, peptides, guanidino compounds, urea, betaines, nucleotides, and quaternary ammonium compounds. The distribution of these compounds varies with taxonomic family (Finne, 1992) as illustrated in Table 3–6. These compounds are important to the seafood technologist because (1) they influence the unique, delicate taste of seafoods and (2) they contribute to the spoilage of fish products.

Table 3–6   Distribution (percentage) of nonprotein-N compounds in fish and shellfish

Class of Compounds	Teleosts (Mackerel)	Elasmobranchs (Shark)	Crustacea (Shrimp)	Molluscs (Squid)
Amino acids	25	5	65	50
Peptides	5	5	15	15
Nucleotides	15	5	5	5
Creatine and creatinine	50	10	—	—
TMAO	15	20	5	15
Urea	—	55	—	—
Betaines	—	—	10	5
Ammonia and amides	5	—	—	—
Octopine	—	—	—	10

Source: Adapted from Finne (1992).

Free Amino Acids

The free amino acid content in muscles of aquatic organisms ranges from about 0.5% to 2% of muscle weight. Free amino acids apparently contribute to osmoregulation (Finne, 1992) and are depleted in fish muscle during starvation (Love, 1988).

Crustacean animals, for example, lobster, shrimp and crab, have a higher content of free amino acids than finfish. Cultured fish muscle tends to have less free amino acids than that of wild fish [reviewed by Haard (1992a)]. Some unique (nonprotein) free acids found in seafood are taurine, sarcosine, β-alanine, methyl-histidine, and α-amino-n-butyric acid. Taurine, a sulfonic acid, is common and abundant in marine invertebrates. In finfish, taurine is normally in greater concentration in white than in red muscle. Taurine can function in osmoregulation, but it may also serve as a food reserve. Taurine is very active in the Maillard browning reaction and contributes to the discoloration of dried squid (Haard and Arcilla, 1985). β-alanine appears to be most common in cold-water fish.

Red muscles tend to contain more free histidine than white muscles. Histidine is the predominant amino acid in fish of the Scombridae family, for example, tuna, mackerel, and mahi mahi. Decarboxylation of histidine by spoilage bacteria gives rise to biologically active histamine. Scombroid poisoning is a form of food intoxication with symptoms similar to allergies discussed under parvalbumins. The allergic reaction is mediated by IgE hypersensitivity which causes release of histamine from mast cells.

Other biologically active amines which are formed in spoiled fish are putrescine and cadaverine (Mietz and Karmas, 1977).

Peptides

Small amounts of peptides have been identified in fish muscle extracts. These include carnosine (β-alanyl histidine), anserine (β-alanyl-1-methylhistidine), and balenine (β-alanyl-3-methylhistidine). Eel muscle is rich in carnosine (500–600 mg%) and dark muscles tend to contain more of these peptides than white muscle. The ratio of carnosine to anserine is normally higher for fresh water than for marine fish. The metabolic role of these peptides is not clear at this time. Carnosine has antioxidant properties, probably due to inhibition of water-soluble lipid oxidation catalysts and/or active oxygen species (Decker and Crum, 1991).

Nucleotides

Normally, more than 90% of the nucleotides in fish muscle are purine derivatives resulting from ATP catabolism. At the moment of death, ATP is the main component of the nucleotide pool. Although ATP catabolism to hypoxanthine in postharvest muscle is generally considered to be the consequence of endogenous enzymes, bacterial inosine nucleosidase can also influence the kinetics of this reaction (Surette, Gill, and LeBlanc, 1980). ATP catabolism occurs at different rates in fish species; for example, ATP is depleted with 1 h after death in free living Pacific rockfish (Lee, Singh, and Haard, 1992) but its depletion may require almost 1 week in farmed sturgeon (Izquerdo-Pulido et al., 1992). Crustacean muscle tends to have low AMP deaminase activity and tends to accumulate AMP rather than IMP immediately after death of the animal.

Other nucleotides in fish muscle include nicotinamide adenine nucleotides (NAD/NADH; 4–38 mg%; NADP/NADPH, 0.3–11 mg%). Normally, dark muscle contains about twice as many of these compounds than white muscle. Ribose phosphates resulting from postharvest catabolism of these compounds contribute to Maillard browning in some seafood products (Haard and Arcilla, 1985).

Guanidino Compounds

The phosphagens, creatine and creatine phosphate, serve as a high-energy reservoir and may be present in muscle at 300–700 mg%. White muscles tend to contain more of these compounds than red muscles.

Table 3–7 Biotoxins found in fish and shellfish

Toxin	Size (daltons)	Structure	Disorder	Sources	Microbes
Saxitoxins	299	$C_{10}H_{17}N_7O_3$	Paralytic shellfish poison	Oysters, clams, scallops	*Protogonyaulax catenella*
Tetrodotoxin	319	$C_{11}H_{17}$ O_8N_3	Pufferfish poison	Pufferfish	*Vibrio alginolyticus*
Domoic acid	311	$C_{15}H_{21}O_6N$	Amnesic shellfish poison	Mussels	*Nitzchia pungens*
Ciguatoxin	1111	$C_{53}H_{77}O_{14}$, $C_{54}H_{78}O_{24}$	Ciguatera	Barracuda, grouper, snapper, amberjack	*Gambierdiscus toxicus*
Histamine	111	$C_5H_9N_3$	Scombroid poison	Tuna, mackerel, mahi mahi	Various bacteria
Brevitoxins	894	$C_{50}H_{70}O_{14}$	Neurotoxic shellfish poison	Oysters, clams, coquina	*Ptychdiscus brevis*
Okadaic acid	804	$C_{44}H_{70}O_{13}$	Diarrhetic shellfish poison	Mussels, scallops	*Dinophysis fortii*

The conversion of creative phosphate to creatine by the Lohman reaction occurs rapidly during stress and slaughter of live animals. Cephalopods, such as squid or scallop, tend to accumulate arginine and octopine, intermediates of the arginine phosphate cycle. Creatinine (6–35 mg%) is formed from creatine by cyclization and appears to be a metabolic end product for excretion.

Urea and Quaternary Ammonium Compounds

Urea is common in all fish tissues but is generally present at less than 50 mg% in muscle of teleost fish. Large amounts of urea are found in marine elasmobranch fish (1–2.5%) where it functions in osmoregulation. Methylamines, like TMAO, appears to stabilize intracellular proteins from denaturation by this protein denaturant (Somero, 1986). Decomposition of urea to ammonia and carbon dioxide is catalyzed by bacterial urease (Finne, 1992). Because urea is water soluble and readily permeates cell membranes, it can be washed from fillets by the processor.

TMAO is common in marine teleosts (15–250 mg%) and present at very high concentrations in marine elasmobranchs (1000–1800 mg%). Red muscles normally contain more TMAO than white muscles from the same specie and concentration can vary with a variety of intraspecific factors. The occurrence and significance of TMAO in marine food products has been reviewed (Hebard, Flick, and Martin, 1982; Haard, 1992c).

Betaines are especially abundant in mollusks and crustacean muscle (400–1000 mg%) where they contribute to taste. The most common betaine is glycine betaine. Other betaines include β-alanine betaine and homarine. Betaine is an intermediate in choline metabolism in the living animal.

Marine Toxins

Seafood may contain nitrogenous compounds that are toxic. The common toxins that are chemically characterized are summarized in Table 3–7. Ciquatera and scombroid poisoning together account for 80% of all seafood outbreaks (Wright, 1992). Several of these toxins are low-molecular-weight, nitrogenous compounds, that is, saxitoxins, domoic acid, ciguatoxin, histamine, and tetrodotoxin. Histamine and tetrodotoxin are produced by bacteria and all others are produced by marine algae.

REFERENCES

AAS, K. 1987. "Fish Allergy and the Codfish Allergen Model. Pp. 356–366. In *Food Allergy and Intolerance*, edited by J. Brostoff. London: Tindall Bailliere.

AMANO, K. 1949. "Studies on the Green Discoloration of Frozen Swordfish (*Xiphias gladius* L.). 1. Isolation of Isovaleric Acid like Substance from Green Colored Flesh." *Bull. Japan. Soc. Sci. Fish.* 15(9):467–474.

ANZAI, H., ASADA, H., KOSHIBA, A., YOSHIDA, S., KOBAYASHI, H., UCHIDA, N., and NISHIDE, E. 1991a. "Distribution of Polysaccharide Digestive Enzymes in a Marine Gastropod *Dolabella auricularia*". *Nippon Suisan Gakkaishi.* 57(11):2077–2081.

ANZAI, H., ENAMI, Y., CHIDA, T., OKOSHI, A., OMURA, T., UCHIDA, N., NISHIDE, E. 1991b. "Polysaccharide Digestive Enzymes from Midgut Gland of Abalone." *Bull. Coll. Agric. Vet. Med., Nihon Univ.* 48:119–128.

ANZAI, H., NANBA, Y., SAWADA, M., UCHIDA, N., and NISHIDE, E. 1992. "Modes of Action of Cellulases from the Gastric Teeth of a Gastropod *Dolabella auricularia*." *Nippon Suisan Gakkaishi.* 58:159–163.

ASSAF, S. A., and GRAVES, D. J. 1969. "Structural and Catalytic Properties of Lobster Muscle Glycogen Phosphorylase." *J. Biol. Chem.* 244:5544–5555.

AUDLEY, M. A., SHETTY, K. J., and KINSELLA, J. E. 1978. "Isolation and Properties of Phospholipase A from Pollock Muscle." *J. Food Sci.* 43:1771–1775.

BELITZ, H.-D., and GROSCH, W. 1987. "Fish, Whales, Crustaceans, Mollusks," p. 462. In *Food Chemistry.* New York: Springer-Verlag.

BERNER, D. L., and HAMMOND, E. G. 1970. "Phylogeny of Lipase Specificity." *Lipids* 5:558–562.

BHUSHANA RAO, K. S. P., FOCANT, B., GERDAY, C., and HAMOIR, G. 1969. "Low Molecular Weight Albumins of Cod White Muscle *Gadus callarias* L." *Comp. Biochem. Physiol.* 30:33–48.

BOBAK, P., and SLECHTA, V. 1988. "Comparison of Parvalbumins of Some Species of Family *Cyprinidae.*" *Comp. Biochem. Physiol.* 91B:697–699.

BRACHO, G. E., and HAARD, N. F. 1991. "Characterization of Alkaline Metalloproteinases with Collagenase Activity from the Muscle of Pacific Rockfish (*Sebastes* sp.)." Pp. 105–125. In *Proceedings of Joint Meeting of Atlantic Fisheries Technologists and Tropical/Subtropical Fisheries Technologists,* edited by S. Otwell. Florida Sea Grant: Gainesville, FL.

BROCKERHOFF, H. 1966. "Digestion of Fat by Cod." *J. Fish. Res. Bd. Canada.* 23:1835–1839.

BROCKERHOFF, H. 1967. "Digestion of Triglycerides by Lobster." *Can. J. Biochem.* 45:421–422.

BROCKERHOFF, H., and HOYLE, J. R. 1965. "Hydrolysis of Triglycerides by the Pancreatic Lipase of a Skate." *Biochim. Biophys. Acta* 98:435–436.

BROOKS, S. P., and STOREY, K. B. 1988a. "Subcellular Enzyme Binding in Glycolytic Control: In Vivo Studies with Fish Muscle." *Amer. J. Physiol.* 255:R289–294.

BROOKS, S. P., and STOREY, K. B. 1988b. "Reevaluation of the 'Glycolytic complex' in Muscle: A Multitechnique Approach Using Trout White Muscle." *Arch. Biochem. Biophys.* 267:13–22.

BROWN, W. D. 1962. "The Concentration of Myoglobin and Hemoglobin in Tuna Flesh." *J. Food Sci.* 27:26–28.

CHAWLA, P., and ABLETT, R. F. 1987. "Detection of Microsomal Phospholipase Activity in Myotomal Tissue of Atlantic Cod (*Gadus morhua*)." *J. Food Sci.* 52(5):1194–1197.

CHEN, S. F. 1993. Effects of dietary vitamin E and fish oil on blood lipoxygenase characteristics and viscosity of cultured grey mullet (*Mugil cephalus*). M.Sc. Thesis, Department of Marine Food Science, National Taiwan Ocean University, 128 pages.

CHOW, C. J., OCHIAI, Y., and HASHIMOTO, K. 1985. "Effect of Freezing and Thawing on the Autooxidation of Bluefin Tuna Myoglobin." *Nippon Suisan Gakkaishi.* 51:2073–2078.

CHOW, C. J., OCHIAI, Y., WATABE, S., and HASHIMOTO, K. 1987. "Autooxidation of Bluefin Tuna Myoglobin Associated with Freezing and Thawing." *J. Food Sci.* 52:589–591, 625.

COWEY, C. B. 1967. "Comparative Studies on the Activity of D-Glyceraldehyde–3-Phosphate Dehydrogenase from Cold and Warm Blooded Animals with Reference to Temperature." *Comp. Biochem. Physiol.* 23:969–976.

CROSTON, C. B. 1965. "Endopeptidases of Salmon Ceca: Chromatographic Separation and Some Properties." *Arch. Biochem. Biophys.* 112:218–223.

DECKER, E. A., and CRUM, A. D. 1991. "Inhibition of Oxidative Rancidity in Salted Ground Pork by Carnosine." *J. Food Sci.* 56:1179–1181.

DE HAEN, C., WALSH, K. A., and NEURATH, H. 1977. "Isolation and Amino-Terminal Sequence Analysis of a New Pancreatic Trypsinogen of the African Lungfish *Protopterus aethiopicus.*" *Biochemistry* 16:4421–4425.

DIXON, M., and WEBB, E. C. 1979. *Enzymes*, 3rd ed. Pp. 633–649. New York: Academic Press.

DUBOIS, I., and GERDAY, Ch. 1990. "Soluble Calcium-Binding Proteins: Parvalbumins and Calmodulin from Eel Skeletal Muscle." *Comp. Biochem. Physiol.* 95B:381–385.

EISEN, A. Z., and JEFFREY, J. J. 1969. "An Extractable Collagenase from Crustacean Hepatopancreas." *Biochem. Biophys. Acta* 191:517–526.

ELSAYED, S., and BENNICH, H. 1975. "Primary Structure of Allergen M from Cod." *Scand. J. Immunol.* 4:203–208.

ESKIN, N. A. M. 1990. *Biochemistry of Foods*, 2nd ed. Pp. 24–26. New York: Academic Press.

FENNEY, R. E. 1988. "Inhibition and Promotion of Freezing: Fish Antifreeze Proteins and Ice-Nucleating Proteins." *Comments Agric. Food Chem.* 1:147–181.

FERRER, O. J., KOBURGER, J. A., SIMPSON, B. K., GLEESON, R. A., and MARSHALL, M. R. 1989. "Phenoloxidase Levels in Florida Spiny Lobster (*Panulirus argus*): Relationship to Season and Molting Stage." *Comp. Biochem. Physiol.* 93B:595–599.

FINNE, G. 1992. "Non-protein Nitrogen Compounds in Fish and Shellfish." Pp. 393–401. In *Advances in Seafood Biochemistry, Composition and Quality*, edited by G. J. Flick and R. E. Martin. Lancaster, PA: Technomic Publishing.

FOLCO, E. J., BUSCONI, L., MARTONE, C. B., and SANCHEZ, J. J. 1989. "Fish Skeletal Muscle Contains a Novel Serine Proteinase with an Unusual Subunit Composition." *Biochem. J.* 263:471–475.

FRANKENNE, F., JOASSIN, L., and GERDAY, C. 1973. "Amino Acid Sequence of the Pike *Esox lucius* Parvalbumin III." *FEBS Lett.* 35:145–147.

FRUTON, J. S., and BERGMANN, M. 1940. "The Specificity of Salmon Pepsin." *J. Biol. Chem.* 136:559–560.

GAZZAZ, S. S., and RASCO, B. A. 1992. "Parvalbumins in Fish and Their Role as Food Allergens: A Review." *Rev. Fish. Sci.* 1:1–26.

GEIST, G. M., and CRAWFORD, D. L. 1974. "Muscle Cathepsins in Three Species of Pacific Sole." *J. Food Sci.* 39:548–551.

GERDAY, C., COLLIN, S., and GERARDIN-OTTHIERS, N. 1989. "Amino Acid Sequence of the Parvalbumin from the Very Fast Swimbladder Muscle of the Toadfish *Opsanus tau.*" *Comp. Biochem. Physiol.* 93B(1):49–55.

GERDAY, C., and TEUWIS, J.-C. 1972. "Isolation and Characterization of the Main Parvalbumins from *Raja clavata* and *Raja montagui* White Muscles." *Biochim. Biophys. Acta* 271:320–331.

GERMAN, J. B., and BERGER, R. 1990. "Formation of 8, 15-Dihydroxy Eicosateraenoic Acid via 15-Lipoxygenase and 12-Lipoxygenase in Fish Gill. *Lipids* 25:849–853.

GERMAN, J. B., and CREVLING, R. K. 1990. "Identification and Characterization of a 15-Lipoxygenase from Fish Gills." *J. Agric. Food Chem.* 38:2144–2147.

GEROMEL, E. J., and MONTGOMERY, M. W. 1980. "Lipase Release from Lysosomes of Rainbow Trout (*Salmo Gairdneri*) Muscle Subjected to Low Temperature." *J. Food Sci.* 45:412.

GHIRETTI, F. 1956. "The Decomposition of Hydrogen Peroxide by Hemocyanin and by Its Dissociation Products." *Arch. Biochem. Phys.* 63:165–176.

GILDBERG. A. 1988. "Aspartic Proteinases in Fishes and Aquatic Invertebrates." *Comp. Biochem. Physiol.* 91B:425–435.

GILDBERG, A., and RAA, J. 1983. "Purification and Characterization of Pepsins from the Arctic Fish Capelin (*Mallotus villosus*). *Comp. Biochem. Physiol.* 75A:337–342.

GRANDJEAN, J., LASZLO, P., and GERDAY, C. 1977. "Sodium Complexation by the Calcium Binding Site of Parvalbumins." *FEBS Lett.* 81:376–380.

GRONINGER, H. S. 1964. "Partial Purification and Some Properties of a Proteinase from Albacore (*Germo alalunga*) Muscle." *Arch. Biochem. Biophys.* 108:175–182.

HAARD, N. F., FELTHAM, L. A. W., HELBIG, N., and SQUIRES, J. 1982. "Modification of Proteins with Enzymes from the Marine Environment." Pp. 223–244. In *Modification of Proteins in Food, Pharmaceutical, and Nutritional Sciences,* edited by R. Feeney and J. Whitaker. *Advances in Chemistry Series* 198. Washington, DC: American Chemical Society.

HAARD, N. F. 1990. "Enzymes from Food Myosystems." *J. Muscle Foods* 1:293–338.

HAARD, N. F. 1992a. "Control of Chemical Composition and Food Quality Attributes of Cultured Fish." *Food Res. Internat.* 25:289–307.

HAARD, N. F. 1992b. "Protein Hydrolysis in Seafood." In *Proceedings World Food Congress,* edited by F. Shahidi and J. R. Botta (*in press*).

HAARD, N. F. 1992c. "Biochemical Reactions in Fish Muscle during Frozen Storage. Pp. 176–209. In *Seafood Science and Technology,* edited by E. G. Bligh. Cambridge, MA: Blackwell Scientific Publications.

HAARD, N. F. 1992d. "Biochemistry and Chemistry of Color and Color Change in Seafoods." Pp. 305–360. In *Advances in Seafood Biochemistry, Composition and Quality,* edited by G. J. Flick and R. E. Martin. Lancaster, PA: Technomic Publishing.

HAARD, N. F., and ARCILLA, R. 1985. "Precursors of Maillard Browning in Dehydrated Squid, *Illex illecebrosus.*" *Can. Inst. Food Sci. Technol. J.* 18:326–331.

HAMEED, K. S., and HAARD, N. F. 1985. "Isolation and Characterization of Cathepsin C from Atlantic Short Finned Squid, *Illex illecebrosus.*" *Comp. Biochem. Physiol.* 82B:241–246.

HAN, T.-J., and LISTON, J. 1987. "Lipid Peroxidation and Phospholipid Hydrolysis in Fish Muscle Microsomes and Frozen Fish. *J. Food Sci.* 52:294–296.

HARA, K., SUZUMATSU, A., and ISHIHARA, T. 1988. "Purification and Characterization of Cathepsin B from Carp Ordinary Muscle." *Nippon Suisan Gakkaishi.* 54:1243–1252.

HAWKINS, D. J., and BRASH, A. R. 1987. "Egg of the Sea Urchin, *Stronglylocentrotus pururatus,* Contain a Prominent (11R) and (12R) Lipoxygenase Activity." *J. Biol. Chem.* 262:762–768.

HEBARD, C. E., FLICK, G. J., and MARTIN, R. E. 1982. "Occurrence and Significance of Trimethylamine Oxide and Its Derivatives in Fish and Shellfish. Pp. 149–304. In *Chemistry and Biochemistry of Marine Food Products,* edited by R. E. Martin, G. J. Flick, C. E. Hebard, and D. R. Ward. New York: Van Nostrand Reinhold.

HENRY, T., and FERGUSON, A. 1985. "Kinetic Studies on the Lactate Dehydrogenase (LDH–5) Isozymes of Brown Trout, *Salmo trutta.*" *Comp. Biochem. Physiol.* 82B:95–98.

HJELMELAND, K., and RAA, J. 1982. "Characteristics of Two Trypsin Type Isozymes Isolated from the Arctic Fish Capelin (*Mallotus villosus*)." *Comp. Biochem. Physiol.* 71B:557–562.

HSIEH, R. J., and KINSELLA, J. E. 1986. "Lipoxygenase-Catalyzed Oxidation of N–6 and N–3 Polyunsaturated Fatty Acids: Relevance to the Activity in Fish Tissue." *J. Food Sci.* 51:940–945.

ILGNER, R. H., and WOODS, A. E. 1985. "Purification, Physical Properties and Kinetics of Peroxidases from Freshwater Crayfish (*Genus orconectes*)." *Comp. Biochem. Physiol.* 82B:433–440.

INABA, T., SHINDO, N., and FUJI, M. 1976. "Purification of Cathepsin B from Squid Liver." *Agric. Biol. Chem.* 40:1159–1165.

IWATA, K., KABASHI, K., and HASE, J. 1973. "Studies on Muscle Alkaline Protease—I. Isolation, Purification and some Physicochemical Properties of an Alkaline Protease from Carp Muscle." *Bull. Japan. Soc. Sci. Fish.* 39:1325–1337.

IWATA, K., KABASHI, K., and HASE, J. 1974. "Studies on Muscle Alkaline Protease—III. Distribution of Alkaline Protease in Muscle of Freshwater Fish, Marine Fish, and in Internal Organs of Carp." *Bull. Japan Soc. Sci. Fish.* 40:201–209.

IZQUERDO-PULIDO, M., HATAE, K., and HAARD, N. F. 1992. "Nucleotide Catabolism and Changes in Texture During Ice Storage of Cultured Sturgeon, *Acipenser transmontanus.*" *J. Food Biochem.* 16:173–192.

JANY, K. D. 1976. "Studies on the Digestive Enzymes of the Stomachless Bonefish *Carassius auratus gibelio* (Bloch): Endopeptidases." *Comp. Biochem. Physiol.* 53B:31–38.

JIANG, S. T., TSAO, C. Y., WANG, Y. T., and CHEN, C. S. 1990. "Purification and Properties of Proteases from Milkfish Muscle (*Chanos chanos*)." *J. Agric. Food Chem.* 38:1458–1463.

JOSEPHSON, D. B., and LINDSEY, R. C. 1986. "Enzymic Generation of Fresh Fish Volatile Aroma Compounds." P. 201. In *Biogenesis of Aromas*, edited by I. H. Parliment and R. Croteau. ACS Symposium No. 317. Washington, DC: American Chemical Society.

KALAC, J. 1978. "Studies on Herring (*Clupea harengus* L) and Capelin (*Mallotus villosus* L) Pyloric Ceca Protease III. Characterization of the Anionic Fractions of Chymotrypsins." *Biologia (Brastislava)* 33:939–945.

KIM, K., and HAARD, N. F. 1992. "Degradation of Proteoglycans in the Skeletal Muscle of Pacific Rockfish (*Sebastles* sp.) During Ice Storage." *J. Muscle Foods* 3:103–121.

KIM, K.-H., and TCHEN, T. T. 1962. "Tyrosinase of the Goldfish *Carassius auratus* L. I. Radio-assay and Properties of the Enzyme." *Biochim. Biophys. Acta* 59:569–576.

KINOSHITA, M., TOYOHARA, H., and SHIMIZU, Y. 1990. "Diverse Distribution of Four Distinct Types of Modori (Gel Degradation)-Inducing Proteinases Among Fish Species." *Nippon Suisan Gakkaishi.* 56:1485–1492.

KINSELLA, J. E., GERMAN, J. B., and SHETTY, J. 1985. "Uricase from Fish Liver: Isolation and Some Properties." *Comp. Biochem. Physiol.* 82B:621–624.

KISHI, H., NOZAWA, H., and SEKI, N. 1991. "Reactivity of Muscle Transglutaminase on Carp Myofibrils and Myosin B." *Nippon Suisan Gakkaishi.* 57:1203–1210.

KITAMIKADO, M., and TACHINO, S. 1960. "Studies on the Digestive Enzymes of Rainbow Trout—II. Proteases." *Bull. Japan Soc. Sci. Fish.* 26:685–694.

KOBUTA, M., OHNUMA, A., and KARUBE, I. 1970. "Kinetics of Protease Inactivation by Gamma Irradiation." *J. Tokyo Univ. Fish.* 55(1):9–20.

KONOSU, S., HAMOIR, G., and PECHERE, J.-F. 1965. "Carp Myogens of White and Red Muscles: Properties and Amino Acid Composition of the Main Low Molecular Weight Components of White Muscle." *Biochem. J.* 96:98–112.

KUO, J.-M., and PAN, B. S. 1992. Occurrence and Properties of 12-Lipoxygenase in the Hamolymph of Shrimp (*Penaeus japonicus* Bate)." *J. Chinese Biochem. Soc.* 21(1):9–16.

LEE, Y.-Z., SINGH, R. P., and HAARD, N. F. 1992. Changes in Freshness of Chilipepper Rockfish (*Sebastes goodei*) During Storage Measured by Biochemical Biosensors." *J. Food Biochem.* 16:119–129.

LEGER, C. 1972. "Essai de purification de la lipase de tissu intercaecal de la truite (*Salmo gairdneri* Rich.)." *Ann. Biol. Biochim. Biophys.* 12:341–345.

LEHKY, P., and STEIN, E. A. 1979. "Perch Muscle Parvalbumin: General Characterization and Magnesium Binding Properties." *Comp. Biochem. Physiol.* 63B:253–259.

LEHRER, S. B., and DAUL, C. B. 1992. "Allergenic Reactions to Seafood: Identification of Allergens." Pp. 185–197. In *Advances in Seafood Biochemistry, Composition and Quality*, edited by G. J. Flick and R. E. Martin. Lancaster, PA: Technomic Publication.

LEHRER, S. B., and McCANTS, M. L. 1987. "Reactivity of IgE Antibodies with Crustacea and Oyster Allergens: Evidence for Common Antigenic Structures." *J. Allergy Clin. Immunol.* 80:133–139.

LOVE, R. M. 1980. *The Chemical Biology of Fishes.* Vol. 2, Pp. 277–278. New York: Academic Press.

LOVE, R. M. 1988. *The Food Fishes: Their Intrinsic Variation and Practical Implications.* P. 59. London: Farrand Press.

LOW, P. S., BADA, J. L., and SOMERO, G. N. 1973. "Temperature Adaptation of Enzymes: Roles of Free Energy, the Enthalpy, and the Entropy of Activation." *Proc. Nat. Acad. Sci.* 70:430–432.

LUNDSTROM, R. C., CORREIA, F. F., and WILHELM, K. A. 1982. "Enzymatic Dimethylamine and Formaldehyde Production in Minced American Plaice and Blackback Flounder Mixed with Red Hake TMAO-ase Active Fraction. *J. Food Sci.* 47:1305.

LUSHCHAK, V. I. 1991. "Characteristics of Microsome-Bound Lactate Dehydrogenase from Skate White Muscles." *Biokhimiia* 56(12):2173–2180.

MACDONALD, R. E., and HULTIN, H. O. 1987. "Some Characteristics of the Enzymic Lipid Peroxidation System in the Microsomal Fraction of Flounder Skeletal Muscle." *J. Food Sci.* 52:15–21.

MEIJER, L., BRASH, A. R., BRYANT, R. W., NG, K., MACLOUF, J., and SPRECHER, H. 1986. "Stereospecific Induction of Starfish Oocyte Maturation by 8(R)-Hydroxyeicosatetraenoic Acid." *J. Biol. Chem.* 261:17040–17047.

MEIJER, L., MACLOUF, J., and BRYANT, R. W. 1986. "Arachidonic Acid Metabolism in Starfish Oocytes." *Dev. Biol.* 114:22–33.

MERRETT, T. G., BAR-ELI, E., and VAN VUNAKIS, H. 1969. "Pepsinogens A, C, and D from the Smooth Dogfish." *Biochemistry* 8:3696–3702.

MIETZ, J. L., and KARMAS, E. 1977. "Chemical Quality Index for Canned Tuna as Determined by High Pressure Liquid Chromatography." *J. Food Sci.* 42:155–158.

MURAKAMI, K., and NODA, M. 1981. "Studies on Proteinases from the Digestive Organs of Sardine I. Purification and Characterization of Three Alkaline Proteinases from the Pyloric Ceca." *Biochim. Biophys. Acta* 658:17–26.

NIMMO, I. A., and SPALDING, C. M. 1985. The Glutathione S-Transferase Activity in the Kidney of Rainbow Trout (*Salmo giardneri*)." *Comp. Biochem. Physiol.* 82B:91–94.

NODA, M., and MURAKAMI, K. 1981. "Studies on Proteinases from the Digestive Organs of Sardines II. Purification and Characterization of Two Acid Proteinases from the Stomach." *Biochim. Biophys. Acta* 658:27–34.

NORRIS, E. R., and ELAM, D. W. 1940. "Preparation and Properties and Crystallization of Tuna Pepsin." *J. Biol. Chem.* 204:673–680.

NUMAKURA, T., SEKI, N., KIMURA, I., TOYODA, K., FUJITA, T., TAKAMA, K., and AKAI, K. 1985. "Cross-linking Reaction of Myosin in Fish Paste During Setting." *Nippon-Suisan Gakkaishi.* 51:1559–1565.

OLLEY, J. PIRIE, R., and WATSON, H. 1962. "Lipase and Phospholipase Activity in Fish Skeletal Muscle and Its Relationship to Protein Denaturation." *J. Sci. Food Agric.* 13:501–516.

OWEN, T. G., and WIGGS, A. J. 1971. "Thermal Compensation in the Stomach of the Brook Trout (*Salvelinus fontinalis* Mitchill)." *Comp. Biochem. Physiol.* 40B:465–473.

PARKIN, K. I., and HULTIN, H. O. 1981. "A Membrane Fraction from Red Hake Muscle Catalyzing the Conversion of TMAO to Dimethylamine and Formaldehyde." *Sci. Technol. Refrig.* 4–6:475–481.

PECHERE, J.-F. 1977. "The Significance of Parvalbumins Among Muscle Calciproteins. Pp. 213–221. In *Calcium-Binding Proteins and Calcium Function*, edited by R. H. Wasserman. New York: North-Holland.

PECHERE, J.-F., DEMAILLE, J., and CAPONY, J. 1971. "Muscular Parvalbumins: Preparative and Analytical Methods of General Applicability." *Biochim. Biophys. Acta* 236:391–408.

PIRONT, A., HAMOIR, G., and CROCAERT, R. 1968. "Proteinic Composition of the Low Ionic Strength Extracts of *Tilapia macrochir* Boulenger White Muscle." *Arch. Internat. Physiol. Biochem.* 76:1–25.

PROTA, G., ORTONNE, J. P., VOULET, C., KHATCHADOURIAN, C., NARDI, G., and PALUMBO, A. 1981. Occurrence and properties of tyrosinase in the ejected ink of cephalopods. *Comp. Biochem. Physiol.* 68B:415–419.

RAMAKRISHNA, M., HULTIN, H. O., and ATALLAH, M. T. 1987. "A Comparison of Dogfish and Bovine Chymotrypsins in Relation to Protein Hydrolysis." *J. Food Sci.* 52(5):1198–1202.

REHBEIN, H., and VAN LESSEN, E. 1989. "Parvalbumin in Salmon and Trout Species." *Arch. Fisch. Wiss.* 39(2):147–162.

RUANO, A. R., RIANO, J. L. A., AMIL, M. R., and SANTOS, M. J. H. 1985. "Some Enzymatic Properties of NAD$^+$-Dependent Glutamate Dehydrogenase of Mussel Hepatopancreas (*Mytilus edulis*)—Requirement of ADP." *Comp. Biochem. Physiol.* 82B:197–202.

SAKAIZUMI, M. 1985. "Species-Specific Expression of Parvalbumins in the Genus *Oryzias* and its Related Species." *Comp. Biochem. Physiol.* 80B:499–505.

SANCHEZ-CHIANG, L., and PONCE, O. 1981. "Gastricsinogens and Gastricsins from

Merluccius gayi—Purification and Properties." *Comp. Biochem. Phys.* 68B:251–257.

SAVAGAON, K. A., and SREENIVASAN, A. 1975. "Purification and Properties of Latent and Isoenzymes of Phenoloxidase in Lobster (*Panulirus homarus* Linn)." *Indian J. Biochem. Physiol.* 12:94–99.

SCHMITT, A., and SIEBERT, G. 1967. "Differentiation of Aliphatic Dipeptidases from Cod Muscle." *Z. Physiolog. Chemi* 348:1009–1016.

SCHWIMMER, S. 1981. *Source Book of Food Enzymology*. P. 498. New York: Van Nostrand Reinhold.

SCOTT, E. M., and HARRINGTON, J. P. 1990. "Comparative Studies of Catalase and Superoxide Dismutase Activity Within Salmon Erythrocytes." *Comp. Biochem. Physiol.* 95B:91–93.

SHEREKAR, S. V., GORE, M. S., and NINJOOR, V. 1988. "Purification and Characterization of Cathepsin B from the Skeletal Muscles of Fresh Water Fish, *Tilapia mossambica*." *J. Food Sci.* 53(4):1018–1023.

SHEWFELT, R. L. 1981. "Fish Muscle Lipolysis—A Review." *J. Food Biochem.* 5:79–100.

SHEWFELT, R. L., McDONALD, R. E., and HULTIN, H. O. 1981. "Effect of Phospholipid Hydrolysis on Lipid Oxidation in Flounder Muscle Microsomes." *J. Food Sci.* 1297–1301.

SIANG, N. C., KWANG, L. H., and CHANG, N. M. 1991. "Some Factors Influencing Gel Strength of Tropical Sardine (*Sardinella gibossa*)." Pp. 236–249. In *Proceeding of Seminar on Advances in Fishery Post-Harvest Technology in Southeast Asia*, edited by H. K. Kuang and M. B. Salim.

SIEBERT, G. 1962. "Enzymes in Marine Fish Muscle and Their Role in Fish Spoilage." Pp. 80–82. In *Fish in Nutrition*, edited by E. Heen and R. Kreuzer. London: London Fishing News.

SIMPSON, B. K., and HAARD, N. F. 1984. "Trypsin from Greenland Cod *Gadus ogac*. Kinetic and Thermodynamic Characteristics." *Can. J. Biochem. Cell Biol.* 62:894–900.

SIMPSON, B. K., and HAARD, N. F. 1987a. "Cold-Adapted Enzymes from Fish." Pp. 495–527. In *Food Biotechnology*, edited by D. Knorr. New York: Marcel Dekker.

SIMPSON, B. K., and HAARD, N. F. 1987b. "Trypsin and Trypsin-like Enzymes from the Stomachless Cunner. Kinetic and Other Physical Properties. *J. Agric. Food Chem.* 35:652–656.

SIMPSON, B. K., MARSHALL, M. R., and OTWELL, W. S. 1988. "Phenoloxidases from Pink and White Shrimp: Kinetic and Other Properties." *J. Food Biochem.* 12:205–217.

SIMPSON, B. K., SMITH, J. P., and HAARD, N. F. 1991. "Marine Enzymes." Pp. 1645–1653. In *Encyclopedia of Food Science & Technology*. New York: Wiley.

SIMPSON, M. V., and HAARD, N. F. 1987. "Temperature Acclimation of Atlantic Cod, *Gadus morhua*, and Its Influence on Freezing Point and Biochemical

Damage of Postmortem Muscle Stored at 0°C and −3°C." *J. Food Biochem.* 11:69–93.

SLABYJ, B. M., and HULTIN, H. O. 1982. "Lipid Peroxidation by Microsomal Fractions Isolated from Light and Dark Muscles of Herring (*Clupea harengus*)." *J. Food Sci.* 47:1395–1398.

SMITH, D. M. 1991. "Factors Influencing Heat-Induced Gelation of Muscle Proteins." Pp. 243–256. In *Interactions of Food Proteins*, edited by N. Parris and R. Barford, American Chemical Society Series 454. Washington DC: American Chemical Society.

SOMERO, G. 1986. "From Dogfish to Dogs: Trimethylamines Protect Proteins from Urea." *NIPS* 1:9–12.

SQUIRES, J., HAARD, N. F., and FELTHAM, L. A. W. 1986. "Pepsin Isozymes from Greenland Cod, *Gadus ogac*. 1. Purification and Physical Properties." *Can. J. Biochem. Cell Biol.* 65:205–209.

STIRLING, W. 1884. "On the Ferments or Enzymes of the Digestive Tract in Fishes." *J. Anat. Physiol.* 18:426–435.

SUMMERS, N. M. 1967. "Cuticle Sclerotization and Blood Phenol Oxidase in the Fiddler Crab, *Uca pugnax*." *Comp. Biochem. Physiol.* 23:129–138.

SURETTE, E., GILL, T. A., and LEBLANC, P. J. 1980. "Biochemical Basis of Nucleotide Catabolism, in Cod (*Gadus morhua*) and Its Relationship to Spoilage." *J. Agric. Food Chem.* 36:19–22.

SUZUKI, T. 1981. *Fish and Krill Protein Processing Technology*. Pp. 10–13. London: Applied Science Publishers.

TANJI, M., KAGEYAMA, T., and TAKAHASHI, K. 1988. "Tuna Pepsinogens and Pepsins. Purification, Characterization and Amino Terminal Sequences." *Eur. J. Biochem.* 177:251–259.

TAPPEL, A. L., SAWANT, P. L., and SHIBKO, S. 1963. Lysosomes: Distribution in animals, hydrolytic capacity and other properties. In *Lysosomes*, edited by D. Reuk and A. Cameron. Pp. 78–108. London: Churchill Ltd.

THEBAULT, M. T., BERNICARD, A., and LENNON, J. F. 1981. "Lactate Dehydrogenase from the Caudal Muscle of the Shrimp *Palaemon serratus:* Purification and Characterization." *Comp. Biochem. Physiol.* 68B:65–70.

TING, C-Y., MONTGOMERY, M., and ANGLEMIER, A. F. 1968. "Partial Purification of Salmon Muscle Cathepsins." *J. Food Sci.* 33:617–621.

TOKUNAGA, T. 1965. "Studies on the Development of Dimethylamine and Formaldehyde in Alaska Pollack. II. Factors Affecting the Formation of Dimethylamine and Formaldehyde." *Bull. Hokkaido Reg. Fish. Res. Lab.* 30:90.

TOOM, P. M., WARD, C. F., and WEBER, J. R. 1982. "Identification of Fish Species by Isoelectric Focusing." Pp. 51–65. In *Chemistry and Biochemistry of Marine Food Products*, edited by R. E. Martin, G. F. Flick, C. E. Hebard and D. R. Ward. New York: Van Nostrand Reinhold.

TSUKUDA, N. 1970. "Studies on the discoloration of red fishes. VI. Partial purification and specificity of the lipoxygenase-like enzyme responsible for carotenoid discoloration in fish skin during refrigerated storage." Nippon Suisan Gakkaishi, 36:725–733.

UENO, R., IKEDA, S., SAKANAKA, K., and HORIGUCHI, Y. 1988. "Characterization of Pepstatin Insensitive Protease in Mackerel Muscle." *Nippon Suisan Gakkaishi* 54:699–707.

WAITE, J. H. 1985. "Catechol Oxidase (EC 1.10.3.1) in the Bysus of the Common Carp *Mytilus edulis.*" *J. Marine Biol. Assoc. U. K.* 65:359–371.

WARREN, G. J., MUELLER, G. M., and McKOWN, R. L. 1992. "Ice Crystal Growth Suppression Polypeptides and Method of Making." U. S. patent 5,118,792, June 2, 1992.

WHITMORE, D. H. 1986. "Identification of Sunfish Species by Muscle Protein Isoelectric Focusing." *Comp. Biochem. Physiol.* 84B:177–180.

WICKS, M. A., and MORGAN, R. P. 1976. "Effects of Salinity on Three Enzymes Involved in Amino Acid Metabolism from the American Oyster, *Crassostrea virginica.*" *Comp. Biochem. Physiol.* 53B:339–343.

WIGGS, A. J. 1974. "Seasonal Changes in the Thyroid Proteinase of a Teleost Fish, the Burbot, *Lota lota* L. *Can. J. Zool.* 52:1071–1078.

WINKLER, M., PILHOFER, G., and GERMAN, J. B. 1992. "Stereochemical Specificity of the N-9-Lipoxygenase of Fish Gill." *J. Food Biochem.* 15:437–448.

WOYCHIK, J. H., and DOWER, H. J. 1990. "Characterization of Parvalbumins of Alaska and Atlantic Pollock and Gulf and Atlantic Menhaden." *IFT National Meeting.* Anaheim, CA. Abstract No. 757. Chicago, IL: Institute of Food Technologists.

WRIGHT, J. L. C. 1992. "Shellfish Toxins: A Canadian Perspective." Pp. 366–375. In *Seafood Science and Technology,* edited by E. G. Bligh. Cambridge, MA: Blackwell Scientific Publications.

YAMADA, A., TAKANO, K., and KAMOI, I. 1991. "Purification and Properties of Amylases from Tilapia Intestine." *Nippon Suisan Gakkaishi.* 57:1903–1909.

YAMAGATA, M., HORIMITO, M., and NAGAOKA, C. 1971. "Accuracy of Predicting Occurrence of Greening in Tuna Based on Content of Trimethylamine Oxide." *J. Food Sci.* 36:55–57.

YOSHINAKA, R., SATO, M., and IKEDA, A. 1978. "Distribution of Collagenase in the Digestive Organs of Some Teleost Fish." *Bull. Japan. Soc. Sci. Fish.* 40(2):263–268.

YOSHINAKA, R., SATO, M., SUZUKI, T., and IKEDA, S. 1984. "Enzymatic Characterization of Anionic Trypsin of the Catfish (*Parasilurus asotus*)." *Comp. Biochem. Physiol.* 77B:1–6.

ZAJICEK, P. 1991. "Enzyme Polymorphism of Freshwater Trypanosomes and Its Use for Strain Identification." *Parasitology.* 102:221–224.

ZEEF, A. H., and DENNISON, C. 1988. "A Novel Cathepsin from the Mussel (*Perna perna*) (Linne)." *Comp. Biochem. Physiol.* 90B:210–204.

4

THE MYOFIBRILLAR PROTEINS
IN SEAFOODS

Zdzisław E. Sikorski

INTRODUCTION

Very concise and informative reviews on the myofibrillar proteins have been published by Maruyama (1985), Squire and Vibert (1987), Nakai and Li-Chan (1988), and Morrissey, Mulvihill, and O'Neill (1987). They describe not only the chemical structure, properties, and location of myosin and actin, of the regulatory proteins, and of the scaffold proteins in the ultrastructure of muscles but also the functional characteristics of several of these proteins.

The myofibrillar proteins present in the muscle in the largest amounts, that is, myosin, actin, tropomyosin, and the troponins C, I, and T, which were first isolated and investigated, are generally best known because they are the main participants in muscle contraction. During the recent two decades, many other components of the myofibrils, which are classified as regulatory proteins and as scaffold proteins, have been isolated and characterized. Although almost each of these proteins presents less than 1% of the total myofibrillar fraction, they may also contribute to the functional properties of meats by interacting with myosin and actin.

Most information on the chemistry and functional properties of the myofibrillar proteins address the proteins of the muscles of rat and of the meat of slaughter animals, mainly beef, mutton, and pork. During the recent decade, however, a large body of research results on fish proteins has been made available in the review by Suzuki (1981). Furthermore, many investigators, interested in problems of comparative biology and in food technology aspects, presented new data on the proteins of fish and marine invertebrates. These data include significant information on changes in the functional properties of fish myofibrillar proteins induced by different parameters of seafood processing.

MYOSIN

Isolation from the Muscles

Different precautions must be applied in extracting myosin from fish meat by salt solutions, to avoid co-extraction of actin and changes in the myosin molecule. In isolating myosin from squid muscle, the contamination of the protein with paramyosin as well as degradation by tissue proteases has to be avoided (Tsuchiya et al. 1978; Kimura et al., 1980). In mammalian muscle, a short-term extraction solubilizes mainly myosin, whereas actin can be extracted only after prolonged treatment. On the other hand, both proteins are being extracted simultaneously as actomyosin from fish muscles. To inhibit the co-extraction of actin and formation of actomyosin, $MgCl_2$ and ATP or pyrophosphate are added to the salt solution. The time of treatment is 6–30 min, according to the properties of different species of fish. Further purification of the myosin solution can be accomplished on DEAE-Sephadex A-50. By using a procedure (Fig. 4-1) proposed by Martone et al. (1986), purified myosin can be isolated from fish muscles in concentrations of 10–15 mg/cm^3 with a yield of 2% of wet tissue.

The procedure of isolation of myosin from muscle tissues may affect the thiol groups in the protein. According to different data the contents of SH residues in myosins from the muscles of 9 species of fish is in the range 29–49 mole/5×10^5 g (Suzuki 1981). In a pure sample of milkfish myosin Chen et al. (1988) found 34 and 38 moles of SH groups per 5×10^5 g protein, as determined by 5,5'-dithiobis-2-nitrobenzoic acid (DTNB) titration and calculated from the amino acid composition data, respectively. However, trout myosin prepared by Buttkus (1970) under conditions excluding all heavy-metal contamination of the sample, con-

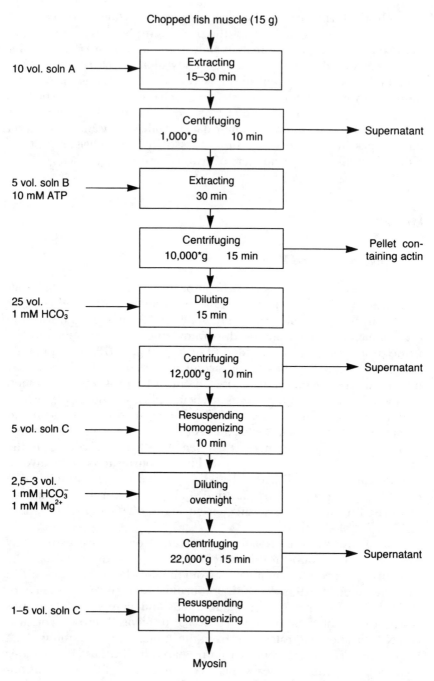

Chopped fish muscle (15 g)

10 vol. soln A → Extracting 15–30 min

Centrifuging 1,000*g   10 min → Supernatant

5 vol. soln B 10 mM ATP → Extracting 30 min

Centrifuging 10,000*g   15 min → Pellet containing actin

25 vol. 1 mM HCO₃⁻ → Diluting 15 min

Centrifuging 12,000*g   10 min → Supernatant

5 vol. soln C → Resuspending Homogenizing 10 min

2,5–3 vol. 1 mM HCO₃⁻ 1 mM Mg²⁺ → Diluting overnight

Centrifuging 22,000*g   15 min → Supernatant

1–5 vol. soln C → Resuspending Homogenizing

Myosin

See caption at bottom of facing page

tained 42 mole SH groups per 5×10^5g protein as determined by DTNB titration but only about 35 moles in the presence of heavy metals.

Subunit Composition

The myosin molecule consists of two heavy chains, which are noncovalently associated with two pairs of light chains. The fish myosins from the ordinary (fast) and dark (slow) skeletal muscle are similar in structure to mammalian and avian myosins. The heavy chains of most myosins exist as isoforms of 200 and 204 kDa. In cardiac muscle, the isoforms form both heterodimers and homodimers, whereas in skeletal muscle, only homodimers are formed. The vertebrate smooth-muscle myosin is also composed of homodimeric heavy chains (Kelley et al., 1992). The amino-terminal part of about 800 residues of each heavy chain folds into a globular domain. Together with one pair of light chains, it composes a globular head. The carboxyl-terminal parts of both heavy chains together form a coiled-coil alpha-helical rod. The globular heads contain the active site for ATPase and the binding site for actin, whereas the helical rod participates in the formation of myosin filaments. The force generation in muscle contraction, triggered by binding Ca^{2+}, is controlled in vertebrate skeletal muscles by the troponin–tropomyosin complex of the actin filaments, whereas in vertebrate smooth muscles and in most invertebrate muscles, the force generation is under control of the light chains of myosin.

There are two types of myosin light chains—the nonessential or DTNB light chains, which dissociate from the myosin molecule after treatment with DTNB, and the essential or alkali light chains (A1 and A2), which dissociate at pH 11. The DTNB chains affect the calcium-binding ability, whereas the alkali light chains have a specific Ca^{2+}-binding site and are essential for the ATPase activity of myosin. The light chains of warm-blooded vertebrate myosin from the fast skeletal muscles are

---

Figure 4–1 A flow sheet of a simple procedure for extraction and purification of fish muscle myosin. Solution A: $0.10M$ KCl, 1 mM phenylmethylsufonylfluoride, 0.02% NaN_3, and 20 mM tris-HCl buffer pH 7.5. Solution B: $0.45 M$ KCl, 5 mM β-mercaptoethanol (β-MCE), 0.2 M Mg $(CH_3COO)_2$, 1 mM ethylene glycol-bis (β-aminoethyl ether) N, N, N′,N′-tetraacetic acid, 20 mM tris-maleate buffer pH 6.8. Solution C: $0.50 M$ KCl, 5 mM β-MCE, 20 mM tris-HCl buffer pH 7.5. Solution D: $0.80 M$ KCl, 5 mM β-MSE, 0.2 M Mg $(CH_3COO)_2$.

Source: Adapted from Martone et al. (1986).

of three distinct kinds when separated by SDS polyacrylamide gel electrophoresis, whereas those from slow skeletal muscles are of two kinds. Similar sets of subunits were isolated from the ordinary and dark muscles of mackerel *Pneumatophorus japonicus* and yellowtail *Seriola quinqueradiata.* The molecular weights of the A1, A2, and DTNB chains of mackerel myosin were 27, 21, and 19 kDa, respectively, and of yellowtail myosin, 28, 16, and 20 kDa, respectively, in the ordinary muscle, whereas those of the dark muscle are 24.5 and 20 kDa and 26 and 20 kDa, respectively (Watabe, Hashimoto, and Watanabe, 1983; Watabe et al., 1983). The molecular weights of the light subunits isolated from mackerel *Scomber scombrus* were 26, 20.5, and 19.2 kDa for the ordinary muscle and 23.6 and 20.2 for the dark muscle (Martinez et al., 1989).

The essential difference between the A1 and A2 light chains in rabbit and chicken myosins is that the A1 chains contain the so-called difference peptide, that is, an additional 41 residues at the N-terminus, rich in alanine, lysine, and proline, with N-terminal trimethylalanine. In the fast skeletal muscle myosin of 16 species of fish, the contents of these amino acids in the A1 chain has also been found to be higher and the isoelectric point more basic than in the A2 chain (Ochiai et al., 1990).

The amino acid sequences of the heavy chains of myosins from a few sources have been deduced from the corresponding nucleotide sequences. The sequence of the myosine heavy chain of the striated adductor muscle of scallop, deduced by Nyitray et al., (1991), is in about 58%, identical to that of other sarcomeric myosins. Recently, the complete sequence of the 1938 amino acid residues of the heavy chain of adult chicken pectoralis muscle myosin has been determined by protein-sequencing techniques, which show the position of the modified residues: N-acetylamine, ε-N-monomethyllysine, ε-N-trimethyllysine, and 3-N-methylhistidine (Maita et al., 1991).

Stability

The fish muscle myofibrillar proteins are thermally less stable than those from mammalian muscles (Connell, 1961). Even less stable are the proteins of squid (Tanaka et al., 1982). Their stability depends not only on the species of fish and the habitat temperature but also on other conditions affecting the state of the proteins in vitro, especially pH and ionic strength. Conditions favorable to aggregation, that is, low pH and ionic strength increase the thermal stability of the protein, and vice versa (Howell et al., 1991). Isolated myosin from cold-water fish is in different aspects much less stable than that of warm-water fish. However, the difference in the thermal melting transition of cod and snapper myosin, which is significant at low ionic strength and pH, that is, in conditions favoring the aggregation of myosin, decreases at higher ionic strength and pH due to lower thermal stability of the snapper myosin in such conditions suitable for monomer formation (Davies et al., 1988). A significant decrease in T_{max} and enthalpy of the thermal transition of myosin was observed by Park and Lanier (1989) in tilapia muscle mince upon the addition of 3% NaCl. It was found, using the surface hydrophobicity probe 8-anilino-1-nafthalenesulfonate and a thermal scanning rheology monitor, that salts belonging to the salting-in type decreased significantly the initial thermal-transition temperature of tilapia myosin (Wicker et al., 1989).

In fish of many different species, living in waters of higher temperature, the Ca^{2+}–ATPase of the myofibrils has significantly higher thermostability than that of fish caught or reared in colder waters (Table 4-1) (Tsuchimoto et al., 1988a, 1988b). The change in the rate constant of inactivation of the myofibrillar ATPase was found to increase with the rearing period of the fish in the respective temperature (Tsuchimoto et al., 1988a). In the fast skeletal muscle of thermally acclimated carp, Watabe et al. (1992) detected myosin isoforms. The thermal acclimation caused changes in the subfragment-1 of myosin and in the size of the light meromyosin. The myosin of fish acclimated at 10°C had higher acto-S1 Mg^{2+}–ATPase activity and lower thermostability than that for carp acclimated at 30°C.

The effect of temperature on the thermal inactivation of purified milkfish myosin Ca^{2+}–ATPase, expressed as the Q_{10} value, is 1.17 between 0 and 10°C and 14.4 in the temperature range 20–30°C (Chen et al., 1988).

Table 4–1 The rate constants K_D of thermal inactivation of myofibrillar Ca^{2+}–ATPase of fishes caught in fishing grounds differing in bottom temperature

No. of Species	No. of Fish	Temperature (°C)	$K_D{}^a$ $(10^{-5}/s)$
32	79	28	20.3
16	26	25	18.6
18	54	17	48.3
4	11	9	63.5
9	33	2	283.0

[a] Mean values.

Source: Adapted from Tsuchimoto et al. (1988b).

Gelation

Fish meat, when ground with salt, forms a viscous sol, which, upon heating, turns into a viscoelastic gel. The rheological characteristics of the gel depend on the properties of the myofibrillar proteins, which are affected by the species and freshness of the fish, as well as on the processing parameters, mainly protein concentration, pH, ionic strength, and temperature (Niwa, 1992). The rate of decline in the strength of gels, resulting from storage of the fish after catch, is a species characteristic (Martinez, 1989). Although gels prepared from Alaska pollack and lizard fish iced 3 days have much lower quality than those from very fresh fish, hoki is a suitable material for surimi even after several days in ice (Macdonald et al., 1990). The gel strength of fish muscle paste can be increased by prolonged ultraviolet irradiation (Taguchi et al., 1989).

In the formation of the network during setting, a major role is played by the aggregation of the myosin heavy chains through different protein–protein interactions. In the subsequent strengthening of the gel, as a result of cooking, disulfide and hydrophobic interactions appear to be involved (Roussel and Cheftel, 1990). It has been established that the heat gelation of myosin is based on an irreversible aggregation of the myosin heads, in which disulfide exchange is involved, and on thermal helix–coil transition of the tail part of the molecules which is followed by formation of a three-dimensional network, buttressed by noncovalent interactions (Samejima et al., 1981; Ishioroshi et al., 1981). At sufficiently high ionic strength, which favors the depolymerization of the myosin filaments and increase in surface hydrophobicity of proteins, even at temperature below 30°C, the hydrophobic bonding between myosin molecules initiates the formation of a gel structure. The development of the gel in the range 30–44°C is attributable to interactions between the tail portions of myosin, whereas the increase in gel elasticity at 51–80°C is mainly due to the involvement of the head portions of the protein. The participation of the hydrophobic amino acid residues of the head portions in the bonding is favored by high ionic strength and by their exposure to the surface due to the temperature effect (Sano et al., 1990a, 1990b). During thermal gelation of carp myosin, the viscous element increases in the temperature ranges 30–41°C and 51–80°C, apparently as a result of the formation of intramolecular cross-linkages, and is accompanied by an increase in the elastic element at an early stage of both ranges of temperature (Sano et al., 1988). The optimum temperature for increasing the breaking strength of salted meat paste from surimi of threadfin bream and for the cross-linking of myosin heavy chains is for threadfin bream and hoki meat in the range 20–30°C and below 30°C, respectively (Lee et al. 1990a, 1990b).

The salted homogenates of muscles from certain fish living in cold-temperate waters, for example, Alaska pollack and hoki, are known for their ability to form gels at temperatures slightly above 0°C in a few hours. This low-temperature setting takes only a few minutes in homogenates of some Antarctic fish, living in very cold waters (Torley et al., 1991). The ability of low-temperature setting apparently reflects the low thermal stability of myosin from these fish.

The gel-forming ability of dark fish meat and myosin pastes has been known to be lower than that of ordinary muscles. This apparently results from the difference in the unfolding abilities of the rods of the myosin and these muscles. The head portions of the heavy chain of myosin from the dark and ordinary muscles do not differ significantly in thermostability, but the ordinary muscle myosin has two thermal transition points, at 36 and 57°C, whereas that of the dark muscle has only one transition point—at 68°C (Lo et al., 1991).

The gelation of different proteins is highly affected by the length of the protein molecule. Enzymatic fragmentation of the myosin molecule decreases significantly the gel-forming ability of the protein. Fish muscles contain a group of serine proteinases of very strong activity against the myosine heavy chain, which are responsible for the modori phenomenon, that is, thermal degradation of fish gels at temperatures above 60°C (see Chapter 7).

The presence of sarcoplasmic proteins may change the rheological properties of the fish gels. The enzymes aldolase and glyceraldehyde-3-phosphate dehydrogenase are difficult to extract at ionic strengths lower than 0.1, as they are strongly bound to the myofibrils. In the muscles of some fish, they constitute more than 40% of the total sarcoplasmic proteins. When left in the myofibrillar fraction, these enzymes precipitate with the myofibrillar proteins during the heat treatment, thus decreasing the strength of the gel. By increasing the ionic strength and pH of the washing solution used in the process of surimi preparation, the contents of these sarcoplasmic proteins can be reduced and the rheological properties of the gel can be improved (Nakagawa and Nagayama, 1989; Nakagawa et al., 1989).

ACTIN

A concise review on the three-dimensional structure of actin and on the involvement of ATP hydrolysis in actin polymerization was recently published by Carlier (1991).

Actin constitutes about 20% of the total amount of myofibrillar proteins in fish muscle. Electrophoretically homogeneous actin, free of tropomyosin and the troponins, was obtained from milkfish by Jiang et al. (1989) by extraction from the acetone powder of the muscles and purification by reversible polymerization and chromatography on Sephadex G-200 and DEAE-Sephadex A-50. The amino acid composition of milkfish actin was found to be similar to that of trout and rabbit actin, except for the higher content of methionine (Table 4-2). The molecule contained five SH groups, with three of them apparently located on the surface. Tsuchiya et al. (1977) obtained a pure actin preparation from the mantle muscle of squid *Ommastrephes sloani pacificus* by extraction of the acetone powder of actomyosin with 2 mM tris-HCl buffer at pH 7.6, containing 0.1 mM ATP. 0.2 mM CaCl$_2$, and 0.5 mM β-mercaptoethanol, and by purification through G–F transformations. This preparation had an amino acid composition similar to that of fish actin (Table 4-2).

The results of Jiang et al. (1989) suggest that actin may increase

Table 4–2   The amino acid composition of actin isolated from different fish muscles

Amino Acid	Number of Residues/Molecule		
	Milkfish[a]	Trout[b]	Squid[c]
Aspartic acid	36.8	36.5	35
Threonine	30.2	29.1	26
Serine	23.0	25.1	27
Glutamic acid	45.2	42.1	41
Glycine	31.3	29.0	33
Alanine	30.2	31.1	32
Valine	22.1	22.3	22
Methionine	15.1	9.8	14
Isoleucine	27.2	26.0	26
Leucine	30.4	30.5	30
Tyrosine	15.3	15.8	15
Phenyloalanine	11.0	12.7	15
Histidine	8.1	9.6	9
Lysine	21.0	23.5	21
Arginine	19.2	20.9	19
Cysteine	4.8	5.0	4
Proline	19.4	21.7	21
Tryptophan	—	3.1	—

Sources: (a) Jiang et al. (1989). (b) Bridgen (1972), cited by Jiang et al. (1989). (c) Tsuchiya et al. (1977).

the thermal stability of myosin in solution, possibly by competition of its SH groups with the myosin SH groups for oxidation.

Several glycolytic enzymes are known to be bound reversibly to F-actin in the muscle structure. The binding depends on the pH, ionic strength, and the presence of certain metabolites and cations. This may affect the extractability of sarcoplasmic proteins. In vitro, a strong binding of fish muscle aldolase and glyceraldehyde-3-phosphate dehydrogenase to F-actin from red sea bream, Pacific mackerel, and carp was shown by Nakagawa and Nagayama (1989).

Fish actin itself, just like bovine muscle actin, does not display the ability to gel upon heating in the presence of NaCl. Carp muscle F-actin, when heated, forms a curd instead of a gel. Thus, it apparently does not contribute in free form to the elasticity of fish gels (Sano et al., 1989e). In comminuted fish products, however, the long thin filaments composed of actin, tropomyosin, and troponin are connected with numerous myosin molecules and, as such, the natural actomyosin participates in gel formation. The gel strength of heated cod muscle actomyosin is affected by the protein concentration, pH, ionic strength, and temperature and time of heating, whereby all these factors are interrelated (Careche et al., 1991). In croaker actomyosin solutions at low concentration, the protein coagulates and precipitates upon heating, without formation of a gel structure. Gelation occurs only at protein concentrations above 5.5%. The rigidity of the gels has two distinct maxima at about 36 and 60°C, the positions of the maxima being at slightly higher temperatures at lower concentrations of the solution (Wu et al., 1985). The elastic component of the viscoelastic gel of natural fish actomyosin increases with heating in the temperature range 32–43°C and, after decreasing sharply between 43 and 52°C, rises again in the range 52–80°C (Sano et al., 1988). The decrease in the elastic element of the gel is proportional to the F-actin/myosin ratio, and F-actin adds to the gel of the natural fish actomyosin the viscous element (Sano et al., 1989a).

The actomyosin from Antarctic krill is characteristic for its very low thermal stability, about 100 times lower than that of Alaska pollack (Nishita et al., 1981).

TROPOMYOSIN

This protein has a significant molecular heterogeneity, as many isoforms of tropomyosin have been described. The number of tissue-specific isoforms in chicken is 11, in human muscle 10, and in horseshoe crab skeletal muscle 7. On the other hand, the amino acid sequences of the homologous isoforms from different species are almost identical (Miya-zaki et al., 1989). Although the molecular weight of the two subunits of vertebrate skeletal muscle tropomyosin is reported to be 33 kDa, the values determined by Tsuchiya et al. (1980) for squid mantle tropomyosin are 35 and 37 kDa. Among the tropomyosins isolated from 25 species of fish, 22 were composed of α-type subunits only, whereas 3 contained α and β-subunits at molar ration 1 : 1 (Seki and Iwabuchi, 1978).

Tropomyosin does not contribute positively to the gel-forming ability of fish meat. Alaska pollack tropomyosin does not set in the presence of NaCl but forms a transparent solution of low viscosity even at 9% w/w. Heating results in the formation of a white precipitate. Adding tropomyosin to natural actomyosin solution or to myosin brings about a significant decrease in the indices of gel properties, that is, gel strength and elasticity in both one-step and two-step heated gels. According to Sano et al. (1989d), this is probably due to inhibiting cross-linking interactions between other myofibrillar proteins involved in gel formation.

TROPONIN

There are several reports available on the properties of troponins isolated from the muscles of fish and marine invertebrates regarding the amino acid composition (Table 4-3), amino acid sequence, conformation of the protein and its changes induced by calcium ions, as well as sensitivity to denaturing agents. In many respects, fish troponin C is very similar to this protein isolated from rabbit muscles (McCubbin et al., 1982) and the troponins isolated from the skeletal muscles of carp. tilapia, big eye tuna, mackerel, and rainbow trout are able to form a functional complex with carp tropomyosin (Seki and Hasegawa, 1978). Troponin C and troponin I from lobster and crayfish muscles were found to exist in several isoforms (Nishita and Ojima, 1990).

PARAMYOSIN

Occurrence and Function in the Muscles

Paramyosin is one of the less generally known muscle proteins, although it has been identified in the mid–1940s. A recent review on the comparative biochemistry of this protein has been published by Kantha et al. (1990). Paramyosin is characteristic for invertebrates and is present

Table 4–3 The amino acid composition of troponin C isolated from different fish and invertebrates

	Number of Residues/Molecule					
Amino Acid	Pike[a]	Dogfish[b]	Hake[c]	Scallop[d]	Lobster[e]	Ascidia[f]
Aspartic acid	24.9	26.9	34.5	16.7	16.9	21.3
Threonine	6.1	8.6	3.1	7.0	5.9	7.4
Serine	7.3	6.7	7.9	6.4	4.4	6.2
Glutamic acid	32.9	39.6	30.6	20.3	17.6	33.4
Glycine	13.0	14.1	10.1	6.8	7.5	10.8
Alanine	10.1	11.3	9.8	7.6	5.5	10.2
Valine	5.5	7.9	6.2	6.1	4.3	8.5
Methionine	8.8	9.0	7.4	3.0	3.3	9.3
Isoleucin	7.8	8.9	8.4	6.2	6.4	8.2
Leucine	14.0	12.7	15.0	13.4	8.7	11.1
Tyrosine	1.9	1.9	1.5	0.9	1.3	2.3
Phenylalanine	9.2	10.7	8.7	5.0	6.0	8.0
Histidine	—	1.0	0.3	—	1.3	2.3
Lysine	8.2	11.4	11.8	13.9	6.7	11.4
Arginine	7.3	5.5	7.2	4.9	3.4	5.2
Half-cystine	1.0	0.8	0.5	—	—	0.9
Proline	1.0	1.6	1.8	—	0.7	2.8
Tryptophan	0.0	1.8	0.0	1.0	—	1.8

Sources: (a) McCubbin et al. (1982). (b) Malencik et al. (1975). (c) Demaille et al. (1974). (d) Sugino et al. (1989). (e) Nishita and Ojima (1990). (f) Takagi and Konishi (1983).

in their muscles in amounts ranging from about 0.1 to 10 times that of myosin (Offer, 1987). It is found in the myofibrillar fraction of the striated muscles in amounts of about 3% in scallop, 14% in squid, 19% in oyster, and 38% in the smooth muscle of oyster (Sano et al., 1986). The ratio myosin/paramyosin in the crude actomyosin fraction from the striated muscles of the bivalve *Aulacomya ater ater* is about 3.8 (Paredi et al., 1990). The filaments of the red adductor muscle of clams and oysters contains 15–30%, and of the white muscle 38–48%, of paramyosin (Kantha et al., 1990). The salt-soluble protein fraction of the foot muscle of abalone contains about 65% of paramyosin (Peyun et al., 1973).

The paramyosin rods form a core of the thick microfibrils of invertebrate muscles which is covered by a layer of myosin. It is assumed that paramyosin has a structural function and affects the orientation of the myosin molecules. It has been suggested that paramyosin is involved in "catch," that is, long-term maintenance of tension in the muscle with little expenditure of energy typical for many molluscan smooth muscles.

Structure

The amino acid composition of paramyosin is characteristic because of large concentrations of amides and basic amino acid residues: Glu 20–23.5%, Asp 12%, Arg 12%, Lys 9%, and an insignificant amount of proline (Kantha et al., 1990).

The molecule has the form of a coiled-coil rod, formed by two different α-helical chains. At both ends of the molecule, there are short nonhelical regions. The molecule contains two pairs of cysteine residues capable of forming disulfide bridges, one at the N-terminus and the other at about one-third of the rod length from the N-terminus. After reduction of the S–S bonds, or denaturation with guanidine hydrochloride, the coiled-coil can be separated into two subunits about 120 nm long. The molecular weight of each of these subunits of paramyosin from different species ranges from 95 to 125 kDa (Kantha et al., 1990). At room temperature, only the C-terminal part of paramyosin is susceptible to proteolysis by trypsin and papain. At pH 2.0, the helical structure of the C-terminal part undergoes deconformation at a transition temperature of 57°C (Kantha et al., 1990; Watabe and Hartshorne, 1990).

Functional Properties in Foods

Paramyosin has a significant effect on the rheological properties of gels prepared from invertebrate meat. Such gels are characteristic in that they are elastic and more cohesive than gels made from fish meat. Paramyosin added to natural actomyosin systems in amounts up to 15% increases the elongation of the gels, and its additions of up to 25% increase the stretching of the cooked product. However, at 50% or more paramyosin, the cooked gels are brittle. Setting of the paramyosin–natural actomyosin system 24 h at 4°C before cooking increases the gel strength (Sano et al., 1986). The positive effect of setting has been observed also in paramyosin–myosin systems, being larger at higher contents of paramyosin (Sano et al., 1989b). The effect of paramyosin on the properties of the gels is greater in systems containing actomyosin than myosin, probably due to inhibition of the dissociation of myosin from actin (Sano et al., 1989c).

CONNECTIN

Connectin has been detected in the white and dark muscles of carp, in the white muscles of oval filefish, mackerel, yellowtail, and Pacific marlin, as well as in the skeletal muscle of the sei whale. In the white muscle of carp it amounts to about 5% of the total muscular proteins, and its content was found to decrease rapidly post-mortem, probably due to increased solubility (Kimura et al., 1981).

REFERENCES

BUTTKUS, H. 1970. "The Sulfhydryl Content of Rabbit and Trout Myosins in Relation to Protein Stability." *Can. J. Biochem.* 49:97–107.

CARECHE, M., CURRALL, J., and MACKIE, I. M. 1991. "A Study of the Effects of Different Factors on the Heat-Induced Gelation of Cod (*Gadus morhua*, L.) Actomyosin Using Response Surface Methodology." *Food Chem.* 42:39–55.

CARLIER, M. F. 1991. "Actin: Protein Structure and Filament Dynamics." *J. Biol. Chem.* 266:1–4.

CHEN, C. S., HWANG, D. C., and JIANG, S. T. 1988. "Purification and Characterization of Milkfish (*Chanos chanos*) myosin." *Nippon Suisan Gakkaishi.* 54:1423–1427.

CONNELL, J. J. 1961. "The Relative Stabilities of the Skeletal–Muscle Myosins of Some Animals." *Biochem. J.* 80:503–509.

DAVIES, J. R., BARDSLEY, R. G., LEDWARD, A. A., and POULTER, R. G. 1988. "Myosin Thermal Stability in Fish Muscle." *J. Sci. Food Agric.* 45:61–68.

DEMAILLE, J., DUTRUGE, E., EISENBERG, E., CAPONY, J. P., and PECHERE, J. F. 1974. "Troponins C from Reptile and Fish Muscles and Their Relation to Muscular Parvalbumins." *FEBS Lett.* 42:173–178.

HOWELL, B. K., MATTHEWS, A. D., and DONNELLY, A. P. 1991. "Thermal Stability of Fish Myofibrils: A Differential Scanning Calorimetric Study." *Internat. J. Food Sci. Technol.* 26:283–295.

ISHIOROSHI, M., SAMEJIMA, K., and YASUI, T. 1981. "Further Studies on the Roles of the Head and Tail Regions of the Myosin Molecule in Heat-Induced Gelation." *J. Food Sci.* 47:114–120, 124.

JIANG, S. T., WANG, F. J., and CHEN, C. S. 1989. "Properties of Actin and Stability of the Actomyosin Reconstituted from Milkfish (*Chanos chanos*) Actin and Myosin." *J. Agric. Food Chem.* 37:1232–1235.

KANTHA, S. S., WATABE, S., and HASHIMOTO, K. 1990. "Comparative Biochemistry of Paramyosin—A Review." *J. Food Biochem.* 14:61–88.

KELLEY, C. A., SELLERS, J. R., GOLDSMITH, P. K., and ADELSTEIN, R. S. 1992. "Smooth Muscle Myosin is Composed of Homodimeric Heavy Chains." *J. Biol. Chem.* 267:2127–2130.

KIMURA, I., YOSHITOMI, B., KONNO, K., and ARAI, K. 1980. "Preparation of Highly Purified Myosin from Mantle Muscle of Squid. *Ommastrephes sloani pacificus.*" *Bull. Japan. Soc. Sci. Fish.* 46:885–92.

KIMURA, S., MIYAKI, T., TAKEMA, Y., and KUBOTA, M. 1981. "Electrophoretic Analysis of Connectin from the Muscles of Aquatic Animals." *Bull. Japan. Soc. Sci. Fish.* 47:787–791.

LEE, N., SEKI, N., KATO, N., NAKAGAWA, N., TERUI, S., and ARAI, K. 1990a. "Gel Forming Ability and Cross-linking Ability of Myosin Heavy Chain in Salted Meat Paste from Threadfin Bream." *Nippon Suisan Gakkaishi.* 56:329–336.

LEE, N., SEKI, N., KATO, N., NAKAGAWA, N., TERUI, S., and ARAI, K. 1990b. "Changes in Myosin Heavy Chain and Gel Forming Ability of Salt-Ground Meat from Hoki." *Nippon Suisan Gakkaishi.* 56:2093–2101.

LO, J. R., MOCHIZUKI, Y., NAGASHIMA, Y., TANAKA, M., ISO, N., and TAGUCHI, T. 1991. "Thermal Transitions of Myosins/Subfragments from Black Marlin (*Makaira mazara*) Ordinary and Dark Muscles." *J. Food Sci.* 56:954–957.

MACDONALD, G. A., LELIEVRE, J., and WILSON, N. D. C. 1990. "Strength of Gels Prepared from Washed and Unwashed Minces of Hoki (*Macruronus novaezelandiae*) Stored in Ice. *J. Food Sci.* 55:976–978.

MAITA, T., YAJIMA, E., NAGATA, S., MIYANISHI, T., NAKAYAMA, S., and MATSUDA, G. 1991. "The Primary Structure of Skeletal Muscle Myosin Heavy Chain: IV. Sequence of the Rod, and the Complete 1,938-Residue Sequence of the Heavy Chain." *J. Biochem.* 110:75–87.

MALENCIK, D. A., HEIZMANN, C. W., and FISCHER, E. H. 1975. "Structural Proteins of Dogfish Skeletal Muscle." *Biochemistry* 14:715–721.

MARTINEZ, I. 1989. "Water Retention Properties and Solubility of the Myofibrillar Proteins: Interrelationships and Possible Value as Indicators of the Gel Strength in Cod Surimi by a Multivariate Data Analysis. *J. Sci. Food Agric.* 46:469–479.

MARTINEZ, I., OLSEN, R. L., OFSTAD, R., JANMOT, C., and D'ALDBIS. A. 1989. "Myosin Isoforms in Mackerel (*Scomber scombrus*) Red and White Muscles." *FEBS Lett.* 252(1,2):69–72.

MARTONE, C. B., BUSCONI, L., FOLCO, E. J., TRUCCO, R. E., and SANCHEZ, J. J. 1986. "A Simplified Myosin Preparation from Marine Fish Species." *J. Food Sci.* 51:1554–1555.

MARUYAMA, K. 1985. "Myofibrillar Cytoskeletal Proteins of Vertebrate Striated Muscle." Pp. 25–50. In *Developments in Meat Science,* Vol. 3, edited by R. Lawrie. London/New York: Elsevier Applied Science.

MCCUBBIN, W. D., OIKAWA, K., SYKES, K., and KAY, C. M. 1982. "Purification and Characterization of Troponin C from Pike Muscle: A Comparative Spectroscopic Study with Rabbit Skeletal Troponin C." *Biochemistry* 21:5948–5956.

MIYAZAKI, J. I., ISIMODA–TAKAGI, T., SEKIGUCHO, K., and HIRABAYASHI, T. 1989. "Comparative Study of Horseshoe Crab Tropomyosin." *Comp. Biochem. Physiol.* 93B:681–687.

MORRISSEY, P. A., MULVIHILL, D. M., and O'NEILL, E. M. 1987. "Functional Properties of Muscle Proteins. Pp. 195–256. In *Developments in Food Proteins,* Vol. 7, edited by B. J. F. Hudson. London/New York: Elsevier Applied Science.

NAKAGAWA, T., and NAGAYAMA, F. 1989. "Interaction of Fish Muscle Glycolytic

Enzymes with F-actin and Actomyosin." *Nippon Suisan Gakkaishi.* 55:165–171.

NAKAGAWA, T., NAGAYAMA, F., OZAKI, H., WATABE, S., and HASHIMOTO, K. 1989. "Effect of Glycolytic Enzymes on the Gel-forming Ability of Fish Muscle." *Nippon Suisan Gakkaishi.* 55:1945.

NAKAI, S., and LI-CHAN, S. 1988. *Hydrophobic Interactions in Food Systems.* Boca Raton, FL: CRC Press.

NISHITA, K., and OJIMA, T. 1990. "American Lobster Troponin." *J. Biochem.* 108:677–683.

NISHITA, K., TAKEDA, Y., and ARAI, K. 1981. "Biochemical Characteristics of Actomyosin from Antarctic Krill." *Bull. Japan. Soc. Sci. Fish.* 47:1237–1244.

NIWA, E. 1992. "Chemistry of Surimi Gelation." Pp. 389–427. In: *Surimi Technology,* edited by T. C. Lanier and C. M. Lee. New York. Basel, Hong Kong: Marcel Dekker.

NYITRAY, L., GOODWIN, E. B., and SZENT-GYORGYI, A. G. 1991. "Complete Primary Structure of a Scallop Striated Muscle Myosin Heavy Chain." *J. Biol. Chem.* 266:18469–18476.

OCHIAI, Y., KOBAYASHI, T., HANDA, A., WATABE, S., and HASHIMOTO, K. 1990. "Possible Presence of the Difference Peptide in Alkali Light Chain 1 of the Fish Fast Skeletal Myosin." *Comp. Biochem. Physiol.* 97B:794–801.

OFFER, G. 1987. "Myosin Filaments." Pp. 307–357. In *Fibrous Protein Structure,* edited by J. M. Squire and Peter J. Vibert. London: Academic Press.

PAREDI, M. E., MATTIO, N. V., and CRUPKIN, M. 1990. "Biochemical Properties of Actomyosin of Cold Stored Striated Muscles of *Aulacomya ater ater* (Molina)." *J. Food Sci.* 55:1567–1570.

PARK, J. W., and LANIER, T. C. 1989. "Scanning Calorimetric Behavior of Tilapia Myosin and Actin Due to Processing of Muscle and Protein Purification." *J. Food Sci.* 54:49–51.

PEYUN, J. H., HASHIMOTO, K., and MATSUURA, F. 1973. "Isolation and Characterization of Abalone Paramyosin." *Bull. Japan. Soc. Sci. Fish.* 39:395–402.

ROUSSEL, H., and CHEFTEL, J. C. 1990. "Mechanisms of Gelation of Sardine Proteins: Influence of Thermal Processing and of Various Additives on the Texture and Protein Solubility of Kamaboko Gels. *Internat. J. Food Sci. Technol.* 25:260–280.

SAMEJIMA, K., ISHIOROSHI, M., and YASUI, T. 1981. "Relative Roles of the Head and Tail Portions of the Molecule in Heat-Induced Gelation of Myosin." *J. Food Sci.* 46:1412–1418.

SANO, T., NOGUCHI, S. F., TSUCHIYA, T., and MATSUMOTO, J. J. 1986. "Contribution of Paramyosin to Marine Meat Gel Characteristics." *J. Food Sci.* 51:946–950.

SANO, T., NOGUCHI, S. F., TSUCHIYA, T., and MATSUMOTO, J. J. 1988. "Dynamic Viscoelastic Behavior of Natural Actomyosin and Myosin During Thermal Gelation." *J. Food Sci.* 53:924–928.

SANO, T., NOGUCHI, S. F., MATSUMOTO, J. J., and TSUCHIYA, T. 1989a. "Role of F-actin in Thermal Gelation of Fish Actomyosin." *J. Food Sci.* 54:800–804.

SANO, T., NOGUCHI, S. F., TSUCHIYA, T., and MATSUMOTO, J. J. 1989b. "Effect of Two-Step Heating on Gel Properties of a Paramyosin–Myosin System." *J. Food Sci.* 54:481–482.

SANO, T., NOGUCHI, S. F., TSUCHIYA, T., and MATSUMOTO, J. J. 1989c. "Paramyosin–Myosin–Actin Interactions in Gel Formation of Invertebrate Muscle." *J. Food Sci.* 54:796–799, 842.

SANO, T., NOGUCHI, S. F., TSUCHIYA, T., and MATSUMOTO, J. J. 1989d. "Contribution of Tropomyosin to Fish Muscle Gel Characteristics." *J. Food Sci.* 54:258–264, 279.

SANO, T., NOGUCHI, S. F., MATSUMOTO, J. J., and TSUCHIYA, T. 1989e. "Dynamic Viscoelastic Behavior of F-actin on Heating. *J. Food Sci.* 54:231–232.

SANO, T., NOGUCHI, S. F., MATSUMOTO, J. J., and TSUCHIYA, T. 1990a. "Effect of Ionic Strength on Dynamic Viscoelastic Behavior of Myosin During Thermal Gelation." *J. Food Sci.* 55:51–54.

SANO, T., NOGUCHI, S. F., MATSUMOTO, J. J., and TSUCHIYA, T. 1990b. "Thermal Gelation Characteristics of Myosin Subfragments." *J. Food Sci.* 55:55–58.

SEKI, N., and HASEGAWA, E. 1978. "Comparative Studies on Fish Troponins." *Bull. Japan. Soc. Sci. Fish.* 44:71–75.

SEKI, N., and IWABUCHI, S. 1978. "On the Subunit Composition of Fish Tropomyosins." *Bull. Japan. Soc. Sci. Fish.* 44:1333–1340.

SQUIRE, J. M., and VIBERT, PETER J. (EDS.). 1987. *Fibrous Protein Structure.* London: Academic Press.

SUGINO, H., KOJIMA, N., NISHITA, K., and OKIMA, T. 1989. "Characterization and Partial Amino Acid Sequence of CNBr-Fragments of Scallop Troponin C." *Nippon Suisan Gakkaishi.* 55:333–340.

SUZUKI, TANEKO. 1981. *Fish and Krill Protein: Processing Technology.* London: Applied Science Publishers.

TAGUCHI, T., ISHIZAKI, S., TANAKA, M., NAGASHIMA, Y., and AMANO, K. 1989. "Effect of Ultraviolet Irradiation on Thermal Gelation of Muscle Pastes." *J. Food Sci.* 54:1438–1440.

TAKAGI, T., and KONISHI, K. 1983. "Amino Acid Sequence of Troponin C Obtained from Ascidian (*Halocynthia roretzi*) Body Wall Muscle." *J. Biochem.* 94:1753–1760.

TANAKA, H., KIMURA, I., ARAI, K., and WATANABE, S. 1982. "The Heat Denaturation of Myosins from Fishes and Squid." *Bull. Japan. Soc. Sci. Fish.* 48:445–453.

TORLEY, P. J., INGRAM, J., YOUNG, O. A., and MEYER-ROCHOW, V. B. 1991. "Salt-Induced, Low-Temperature Setting of Antarctic Fish Muscle Proteins." *J. Food Sci.* 56:251–252.

TSUCHIYA, T., SHINOHARA, T., and MATSUMOTO, J. J. 1980. "Physico-chemical Properties of Squid Tropomyosin." *Bull. Japan. Soc. Sci. Fish.* 46:893–896.

TSUCHIYA, T., SUZUKI, H., and MATSUMOTO, J. J. 1977. "Isolation and Purification of Squid Actin." *Bull. Japan. Soc. Sci. Fish.* 43:1233–40.

TSUCHIYA, T., YAMADA, N., and MATSUMOTO, J. J. 1978. "Extraction and Purification of Squid Myosin." *Bull. Japan. Soc. Sci. Fish.* 44:175–179.

TSUCHIMOTO, M., TANAKA, N., UESUGI, Y., MISIMA, T., TACHIBANA, K., YADA, S., SENTA, T., and YASUDA, M. 1988a. "The Influence of Rearing Water Temperature on the Relative Thermostability of Myofibrillar Ca^{2+}–ATPase and on the Lowering Speed of Freshness in Carp." *Nippon Suisan Gakkaishi.* 54:117–122.

TSUCHIMOTO, M., TANAKA, N., MISIMA, T., YADA, S., SENTA, T., and YASUDA, M. 1988b. "The Influence of Habitat Water Temperature on the Relative Thermostability of Myofibrillar Ca^{2+}–ATPase in Fishes Collected in the Waters from Tropical to Frigid Zones." *Nippon Suisan Gakkaishi.* 54:787–793.

WATABE, S., DIHN, T. N. L., OCHIAI, Y., and HASHIMOTO, K. 1983. "Immunochemical Specificity of Myosin Light Chains from Mackerel Ordinary and Dark Muscles." *J. Biochem.* 94:1409–1419.

WATABE, S., and HARTSHORNE, D. J. 1990. "Paramyosin and the Catch Mechanism." *Comp. Biochem. Physiol.* 96B:639–646.

WATABE, S., HASHIMOTO, K., and WATANABE, S. 1983. "The pH Dependency of Myosin ATPases from Yellowtail Ordinary and Dark Muscles." *J. Biochem.* 94:1867–1875.

WATABE, S., HWANG, G. C., NAKAYA, M., GUO, X. F., and OKAMOTO, Y. 1992. "Fast Skeletal Myosin Isoforms in Thermally Acclimated Carp." *J. Biochem.* 111:113–122.

WICKER, L., LANIER, T. C., KNOPP, J. A., and HAMANN, D. D. 1989. "Influence of Various Salts on Heat-Induced ANS Fluorescence and Gel Rigidity Development of Tilapia (*Serotherodon aureus*) Myosin." *J. Agric. Food Chem.* 37:18–22.

WU, M., C., LANIER, T. C., and HAMANN, D. D. 1985. "Rigidity and Viscosity Changes of Croacker Actomyosin During Thermal Gelation." *J. Food Sci.* 50:14–19, 25.

5

COLLAGEN IN THE MUSCLES AND SKIN OF MARINE ANIMALS

Zdzisław E. Sikorski and Javier A. Borderias

LOCATION OF COLLAGEN AND CONNECTIVE TISSUE IN THE MUSCLES

Collagen is the major protein of the connective tissues. These tissues are composed of a highly hydrated, amorphous ground substance in which several types of cells, as well as the fibrous proteins collagen and elastin, are embedded. The cells fulfill many metabolic functions, including the synthesis of collagen and elastin. The ground substance is composed mainly of water, of proteoglycans which are composed of a protein core with covalently bound glycosamin glycans with sulphate and carboxylate groups, and of glycoproteins. The components of the ground substance interact with collagen and affect the collagen structures. The fibrous proteins form fibrous, filamentous, or network structures, depending on the kind of tissue. The mechanical properties of the connective tissues depend on the size, orientation, and cross-linking of the collagen fibrils and of elastic fibers, as well as on the proportion of all components of the tissue, including the mineral deposits. The biosynthesis and chemistry of collagen has been reviewed by Bailey and Etherington (1980), and its role as a food component has been thoroughly treated by Bailey and

Light (1989) and Bremner (1992), as well as in the book edited by Pearson et al. (1985).

In fish muscles, the collagen fibrils are in the endomysium and perimysium, that is, in membranes enclosing each muscle fiber and muscle bundle, respectively, and in the myocommata, that is, connective tissue sheets separating the adjacent blocks of muscle fibers. These muscle blocks are called myotomes. The collagen fibrils in the fish endomysium are smaller in diameter than the fibrils in the myocommata. The fish muscle fibers are short and run along the long axis of the body. The junction between the muscle fibers and the connective tissue sheets is provided by collagenous microfilaments which connect the collagen fibers of the myocomma with the basal lamina and the sarcolemma in the invagination near the base of the muscle fiber (Fig. 5-1). The thickness of the myocommata depends on the species of the fish, as well as on the age and feeding status of the specimen.

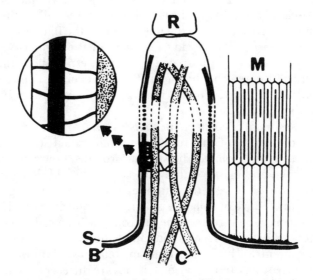

Figure 5–1 Diagrammatic view of the muscle fiber base region in prerigor fish. The invagination is lined by the sarcolemma and a basal lamina. Microfilaments connect the sarcolemma, basal lamina, and collagen fibers (detailed in the inset). Both basal lamina (B) and collagen fibers (C) terminate near the end of the invagination, which is closely appressed to the sarcoplasmic reticulum (R). Myofibrils (M) are connected to the sarcolemma of the muscle fiber base by fine filaments, which run into a darkly stained, amorphous region close to the sarcolemma (S).

Source: Hallett and Bremner (1988).

In the muscles of squid, the block of muscle fibers made of circumferential bands sandwiched between thinner radial bands, is covered on each side by two connective tissue sheets, as has been clearly presented by Otwell and Giddings (1980).

The connective tissues are responsible for the integrity of the fish fillet and for the rheological characteristics of seafoods. The chemical and physical properties of collagen of marine fish and invertebrates, as well as the role of the connective tissues in quality changes of seafoods due to storage and processing, has been reviewed by Sikorski et al. (1984).

THE CONTENTS OF COLLAGEN IN THE TISSUES OF FISH AND SQUID

According to early data, the muscles of bony and cartilaginous fishes contain 2–5% and 11% connective tissue, respectively (Jacquot, 1961). As the muscle fibers decrease in length toward the tail of the fish, the amount of connective tissue in the tail section of the fillet is up to several times larger than near the head. The yield of the connective tissue of fish may increase during the spawning season, especially if the animals are in a depleted state. The thickness of cod myocommata increases with the age of the fish; furthermore, it is larger in winter than in summer (Love, 1980).

The assessment of collagen in the tissues of marine animals is usually based on the quantity of the insoluble protein called stroma or on the contents of hydroxyproline. The stroma proteins, however, consist not only of collagen and elastin but also of connectin and denatured other muscle proteins. Thus, very thorough separation of noncollagenous material in the procedure of collagen isolation is required. Significant losses of soluble forms of collagen may be unavoidable, especially in extracting noncollagenous proteins from squid muscle homogenates. On the other hand, the contents of hydroxyproline in collagens of various species of fish and invertebrates and of various tissues are different, so that specific conversion factors should be used for calculating the amount of collagen in different sources. The conversion factor from hydroxyproline to collagen in the muscles, skin, and membranes of the squid *Illex argentinus* is 22.72, 18.87, and 12.50, respectively but for *Loligo patagonica,* it is 20.41, 16.39, and 13.51, respectively (Sadowska and Sikorski, 1987). According to Montero et al. (1990), the conversion factor for the muscle collagen of hake (*Merluccius merluccius*) and trout (*Salmo irideus*) is 13.3 and 17.8, respectively, whereas for skin collagen, it is 14.4 and 14.2, respectively. In tissues containing free, noncollagenous hydroxyproline and collagen of unknown content of hydroxyproline, the collagen should be determined by isolation of soluble and insoluble collagens, for example, by the simplified method of Sato et al. (1986a), followed by nitrogen determination.

About 3% of the proteins in the muscles of bony fish and about 10% in elasmobranchs constitute the stroma (Dyer and Dingle, 1961). The total collagen in the ordinary muscle in the dorsal part of the trunk of fishes belonging to 24 species, determined by Sato et al. (1986b), was about 0.3–2.2% wet tissue and 1.6–12.4% crude protein. The highest values were found in Japanese eel and conger eel and generally in the musculature of fish with a more flexible body. Furthermore, the ordinary muscles of dark-fleshed fishes contained more collagen than the corresponding muscles from white fishes. The white carp muscle contains about 2.4% collagen, the abdominal muscle tissues of giant river prawn, fleshy prawn, and spiny lobster 2.4–2.6%, the octopus arm muscle about 7%, and the muscular tissue of the foot of top shell about 33% collagen in total tissue protein (Kimura and Tanaka, 1983, 1986). The collagen to crude protein ratio in the visceral, opercular, and foot muscle of turban shell is 5, 34, and 45, respectively (Ochiai et al., 1985). The collagen content in the muscles of fish may vary also with the muscle part, the age of the animal, the season, and the nutritional condition. In hake, the contents of collagen may be about 0.2% near the head and 0.9% in the tail portion of the fillet (Montero and Borderias, 1989b). According to Yoshinaka et al. (1988), the distribution pattern of collagen in the body muscles is closely related to the swimming mode of the fish. The flexible body musculature contains more collagen than other muscles. In starving fish, the sarcoplasmic and myofibrillar proteins undergo gradual depletion, whereas the connective tissues are not utilized, or extra collagen is even deposited in the myocommata and in the skin (Love, 1970).

The skin of hake and trout contains 33.5 and 33.3% collagen on a wet basis (Montero, 1988).

CHEMICAL CHARACTERISTICS OF SEAFOOD COLLAGEN

Type and Subunit Composition

The main collagen component in the muscles and skin of marine fish, lobster, and octopus is type I collagen, which is accompanied by smaller amounts of type III. Sato et al. (1991) demonstrated the presence of type I and type V collagens in the muscles of fish of many species. There are indications that although the total content of type V collagen in the muscle of fish is much lower than that of type I, it may be the major constituent of the endomysium and perimysium. According to Kimura et al. (1988), the type I collagen of skin and of the ordinary muscle of eel (*Anguilla japonica*) and common mackerel (*Scomber japonicus*) has a subunit composition of $\alpha 1\alpha 2\alpha 3$ and that of saury (*Cololabis saira*) $(\alpha 1)_2\alpha 2$, and the collagen of carp (*Cyprinus carpio*) is characteristic of having a major component $(\alpha 1)_2\alpha 2$ and a minor one, composed of $\alpha 1\alpha 2\alpha 3$. The subunit composition of muscle type I collagen of chum salmon (*Onchorhynchus keta*) is identical to that of carp, but that of the skin is $\alpha 1\alpha 2\alpha 3$. Also, the collagen from the skin of blue grenadier (*Macruronus novaezelandiae*) is composed of three different a chains (Ramshaw et al., 1988). The muscle collagens of the top shell *Turbo cornutus* and of the abalone have a structure of $(\alpha 1)_3$, whereas the collagen from the arm muscle of octopus has a subunit composition of $(\alpha 1)_2\alpha 2$ (Takema and Kimura, 1982).

Amino Acid Composition and Contents of Saccharides

Earlier results on the amino acid composition of the muscle collagen of marine fish and invertebrates and on the contents of different saccharides in the collagens were reviewed by Sikorski et al. (1984). The amino acid composition of edible marine fish, molluscan, and crustacean collagens differs from that of mammalian intramuscular collagen in being richer in several essential amino acids and significantly poorer in hydroxyproline. The composition of isolated type I collagen from the muscles of several teleost fishes has been determined by Kimura et al. (1988) (Table 5-1). The composition of type I collagen from the skin, phasciae, myocommata, and myotomes of hake and trout was presented recently by Montero et al. (1990). The top shell muscle collagen characterized by Kimura and Tanaka (1983) contained significantly more aspartic acid, glutamic acid, and hydroxylysine than the teleost muscle collagen.

With respect to the degree of glycosylation, the fish collagens do not differ significantly from mammalian collagen, whereas the collagens of edible marine invertebrates contain several times more carbohydrates, mainly as glucosylgalactosylhydroxylysine residues.

FUNCTIONAL PROPERTIES OF SEAFOOD COLLAGEN

Solubility

The physical properties of fish muscle collagens, that is, the solubility and the hydrothermal shrinking as well as the breaking strength of

Table 5–1 Amino acid composition of type I collagen from the ordinary muscle of teleost fishes (residues/1000)

Amino Acid	Carp	Eel	Mackerel	Saury	Chum Salmon
Hydroxyproline	84	74	72	70	63
Aspartic acid	47	44	48	48	48
Threonine	26	22	27	22	20
Serine	35	32	44	47	48
Glutamic acid	72	69	73	68	74
Proline	110	110	103	103	103
Glycine	337	347	334	350	363
Alanine	119	133	123	115	113
Valine	20	17	19	18	17
Methionine	13	15	14	14	15
Isoleucine	10	10	11	11	10
Leucine	22	20	24	24	19
Tyrosine	3	2	3	4	1
Phenylalanine	13	14	14	14	17
Hydroxylysine	7	6	7	8	9
Lysine	25	24	26	26	23
Histidine	5	10	6	6	7
Arginine	52	51	52	52	50
Totals	1000	1000	1000	1000	1000

Source: Kimura et al. (1988).

the fish connective tissue have been discussed by Sikorski et al. (1984). During the recent decade, several publications contributed new data.

The solubility of fish collagen is generally significantly higher than that of ovine, bovine, or porcine muscles but is not so much affected by the age of the animal, as is known for mammalian collagen. However, the situation is obscured by the thickening of the myocommata and variation in intermolecular cross-linking of myocommata collagen during the periods of seasonal starvation and depletion of the muscles due to sexual maturation. Fish collagens contain the same cross-links which are known to exist in mammal collagens. However, the contents of lysinonorleucine, hydroxylysinonorleucine, and dihydroxylysinonorleucine residues, responsible for the reducible cross-links, in collagen isolated from the muscles of two species of rockfish were found to be different (Bracho and Haard, 1990). Montero and Borderias (1990b) found that the acid-soluble fraction of muscle collagen of trout was significantly higher in older than in younger animals, which was accompanied by a higher thermal solubility of the collagen from the oldest specimens.

The solubility of collagen from the ordinary muscle of the dorsal part of fishes belonging to 24 species examined by Sato et al. (1986b) ranged from 23 to 73%, but the mean value for 19 species was 57%. The data in Table 5-2 should be regarded as representing approximate values only, as they are affected by the age of the animals, postmortem storage, muscle characteristics, and differences in the procedure of solubility determinations. The solubility of muscle collagen of edible marine fish and squid is generally lower than that of the skin collagen.

Different effects of refrigerated storage of fish and squid muscle on the solubility of collagen have been reported. The results are inconsistent because various experimental procedures, affecting the activity of endogenous lysosomal enzymes as well as the accumulation of lactic acid, were used. Furthermore, the susceptibility of fish of various species to gaping due to changes in the muscle connective tissues is known to be very different. Montero and Borderias (1990a) found significant increase in collagen solubility in trout muscle homogenates, kept 3 days under refrigeration, both in raw and cooked samples. On the other hand, no difference in collagen solubility was found in trout and Japanese mackerel analyzed fresh and after 5 days in ice (Sato et al., 1986a). However, in a recent article, Sato et al. (1991) demonstrated a significant increase in solubility of type V collagen in rainbow trout muscle stored 1 day at 5°C. After 3 days, the contents of insoluble collagen of type I decreased by about 50% and of type V by 75%. The increase in the solubility of type V collagen was accompanied by a softening of the muscle. Fish muscles contain proteinases capable of degrading nonhelical regions of collagen. The cleaving of these regions or/and of cross-links of type V collagen may cause softening of the fish muscle during storage. Scanning

Table 5–2   The solubility of collagen from the muscles (M) and skin (S) of fish

Species	Solubility (% of Total Collagen)						Source
	M	S	M	S	M	S	
Hake	5.1	5.2	54.9	60.0	39.9	34.8	1
	4.8	2.5	69.6	90.6	25.6	6.9	2
Cod	4.2	1.5	76.3	85.9	19.5	12.6	1
Catfish	8.3	2.3	42.0	62.6	49.7	35.1	1
Trout	2.6	2.6	70.4	93.9	27.0	3.5	2
	3.8	—	69.2	—	27.0	—	3
Blue grenadier	—	4.8	—	89.8	—	5.5	4

Sources: (1) Yamaguchi et al. (1976). (2) Montero et al. (1990). (3) Montero and Borderias (1990b). (4) Ramshaw et al. (1988).

electron microscopy investigations of muscles from blue grenadier, which is known to be prone to gaping after rigor mortis, revealed a progressive breakdown of the collagen junctions between the myocommata and the muscle fibers during storage in ice. These changes were apparently due to the activity of collagenases and/or other proteinases (Bremner and Hallet, 1985; Hallett and Bremner, 1988).

Frozen storage of hake muscle brings about a significant decrease in solubility of collagen, especially pronounced in minced tissue kept at $-12°C$, which is accompanied by a corresponding accumulation of dimethylamine (DMA) and formaldehyde (Montero and Borderias, 1989a, 1989c). Gill et al. (1991) reported that the collagen solubility in neutral salt solutions was reduced after frozen storage of the fish. The changes in solubility and the electrophoretic evidence suggested collagen cross-linking or interactions of collagen with other cellular proteins during storage. Scanning electron microscopy revealed that collagen fibrils became thickened and the porous nature of the cell surface disappeared following frozen storage.

Thermal Denaturation

The hydrothermal shrinking of collagen fibers and the thermal denaturation of collagen in solution are important characteristics affecting the quality of a variety of fish products. The shrinkage temperature of fish skin collagen is generally significantly lower than that of mammalian hide collagens. It is directly related to the hydroxyproline content (Takahashi and Tokoyama, 1954). In codfish skin collagen, containing 7.7% hydroxyproline, the shrinkage temperature is 40°C, but it is 55–57° C in wild goldfish collagen, containing 12.1% hydroxyproline. The thermal denaturation temperature of fish collagens increases with the contents of hydroxyproline and proline (Yamaguchi et al., 1976). The denaturation temperature of the dark muscle collagen of skipjack (*Euthynnus pelamis*) was found to be higher than that of collagen from the ordinary muscle and skin (Zhu and Kimura, 1991). The breaking strength of fish myocommata decreases significantly with the increase in temperature and decrease in pH (Love et al., 1972). The temperature at which thermal denaturation begins and the total enthalpy of denaturation of bovine intramuscular collagen decreases after treatment with collagenase, the change depending on the degree of cross-linking of the protein (Beltran et al., 1991).

The technological implications of thermal changes in collagen in fish, squid, and other molluscs cookery, in hot smoking of fish, in canning, in the formation of the rheological properties of comminuted cooked products, and in manufacturing of fish meal, silage, and hydrolyzed fish products have been discussed by Sikorski et al. (1984).

Other Functional Properties

The fish skin and muscle collagen has the ability to form gels after cooking. This gel-forming ability depends on the species characteristics of the fish and is significantly higher in the skin collagenous material than in that obtained from the muscles. The addition of NaCl decreases significantly the strength of the gels prepared from the skins but has a lesser effect on the gel-forming ability of the muscle collagenous material (Montero and Borderias, 1990c).

The water-holding capacity, viscosity, and emulsifying capacity of collagenous material from the muscle connective tissue and skin of fish is highly affected by pH and the concentration of NaCl (Montero and Borderias, 1991; Montero et al., 1991).

THE EFFECT OF COLLAGEN ON THE QUALITY OF SEAFOODS

The overall quality of seafoods, that is, the characteristic properties which determine the degree of excellence, comprises both the wholesomeness and the sensory acceptability of the products by the consumers. One of the main factors affecting wholesomeness is the nutritional value of the proteins contained in the products. Collagen, which is comparatively poor in essential amino acids, adds little to the nutritional value of fishery products. However, it affects significantly the tensile strength and integrity of the muscles and the rheological properties of the muscles and fishery products.

The sensory quality of seafoods is determined mainly by flavor and texture, both attributes having a different impact on the overall sensory preference in fish of different species and in different periods of storage after catch. In fish of prime freshness, belonging to the most valued species, the decisive factor, responsible for the high quality, may be the delicate texture of the flesh in addition to the flavor. In other species of fish or squid, or in frozen fish after prolonged storage, the toughness of the product may be the reason for the low-quality note. The most valuable species of fish, for example, the sturgeon, salmon, some flatfish, yellowfin tuna, big eye tuna, and snapper, have very high eating quality because of their delicate texture and flavor.

Although the texture of fish and marine invertebrates is, within a species, highly affected by age, season, nutritional state, and postmortem treatment, there is a significant effect of collagen, both on the raw and cooked texture. The raw muscles of sardine, brook masu salmon, argentine, rainbow trout, and horse mackerel, containing 1.6–2.3% collagen, are tender, whereas Japanese eel, pink conger, and conger eel, with 8.8–12.4% collagen, are very tough (Sato et al. 1986). A positive correlation between the collagen content and the toughness of raw muscles of turban shell was noted by Ochiai et al. (1985). On the other hand, fish containing low amounts of collagen tend to be dry and fibrous after cooking, whereas Japanese eel, pike conger, and conger eel make a very tender, succulent, and elastic product after cooking, possibly due to gelatinization of collagen. Conditioning of squid mantle prior to cooking, favoring the gelatinization of the muscle collagen, decreases the toughness of the cooked product (Kołodziejska et al., 1987). The investigations of Kuo et al. (1991) indicate that the tensile properties of squid mantle depend on the contribution of collagen and of the contractile elements. Although treatment of the mantle with collagenase reduced the tensile strength longitudinally, not affecting it in the transverse direction of the raw sample, soaking in sodium chloride solution and trypsin reduced the strength only transversely. All factors were effective in both directions when the samples were heated after treatment.

Changes in the properties of collagen are involved in the phenomenon of gaping of fish fillets, when the connective tissues fail to hold the muscle blocks together. The direct cause of gaping is the rupturing of the connection of the myocomma with the basal lamina and the sarcolemma near the base of the muscle fiber. Gaping was thoroughly studied by Love and co-workers, who described the effect of species and size of the fish, of the nutritional and biochemical condition, of temperature, and of conditions of handling postmortem. In a recent paper of the series, Love investigated gaping in farmed fish (Lavety et al., 1988).

REFERENCES

BAILEY, A. J., and ETHERINGTON, D. J. 1980. "Metabolism of Collagen and Elastin." Pp. 299–408. In *Comprehensive Biochemistry*, edited by M. Florkin M., A. Neuberger, and L. L. M. van Deenen. New York: Elsevier Applied Science.

BAILEY, A. J., and LIGHT, N. D. 1989. *Connective Tissue in Meat and Meat Products.* London/New York: Elsevier Applied Science.

BELTRAN, J. A., BONNET, M., and OUALI, A. 1991. "Collagenase Effect on Thermal Denaturation of Intramuscular Collagen." *J. Food Sci.* 56:1497–1499.

BRACHO, G. E., and HAARD, N. F. 1990. "Determination of Collagen Crosslinks in Rockfish Skeletal Muscle." *J. Food Biochem.* 14:435–451.

BREMNER, H. A. 1992. "Fish Flesh Structure and the Role of Collagen—its Post-mortem Aspects and Implications for Fish Processing." Pp. 39–62. In *Quality Assurance in the Fish Industry*, edited by H. H. Huss, M. Jakobsen, and J. Liston. London/New York: Elsevier Science Publishers.

BREMNER, H. A., and HALLETT, I. C. 1985. "Muscle Fiber–Connective Tissue Junctions in the Fish Blue Grenadier (*Macruronus novaezelandiae*). A Scanning Electron Microscope Study." *J. Food Sci.* 50:975–980.

DYER, W., and DINGLE, J. R. 1961. "Fish Proteins with Special Reference to Freezing." Pp. 275–327. In *Fish as Food*, Vol. 1, edited by Georg Borgstrom, New York/London: Academic Press.

GILL, T., WALTON, C., and ODENSE, P. 1991. "Changes in Cod (*Gadus morhua*) Collagen in Frozen Storage." Paper read at Diamond Jubilee Conference, August 1991, Lingby, Denmark.

HALLETT, I. C., and BREMMER, H. A. 1988. "Fine Structure of the Myocommata–Muscle Fibre Junction in hoki (*Macruronus novaezelandiae*)." *J. Sci. Food Agric.* 44:245–261.

JACQUOT, R. 1961. "Organic Constituents of Fish and Other Aquatic Animal Foods." Pp. 145–209. In *Fish as Food*, Vol. 1, edited by Georg Borgstrom, New York/London: Academic Press.

KIMURA, S., and TANAKA, H. 1983. "Characterization of Top Shell Muscle Collagen Comprising Three Identical α Chains." *Bull. Japan. Soc. Sci. Fish.* 49:229–232.

KIMURA, S., and TANAKA, H. 1986. "Partial Characterization of Muscle Collagens from Prawns and Lobster." *J. Food Sci.* 51:330–332, 339.

KIMURA, S., ZHU, X. P., MATSUI, R., SHIJOH, M., and TAKAMIZAWA, S. 1988. "Characterization of Fish Muscle Type I Collagen." *J. Food Sci.* 53:1315–1318.

KOŁODZIEJSKA, I., SIKORSKI, Z. E., and SADOWSKA, M. 1987. "Texture of Cooked Mantle of Squid *Illex argentinus* as Influenced by Specimen Characteristics and Treatments." *J. Food Sci.* 52:932–935.

KUO, J. D., HULTIN, H. O., ATALLAH, M. T., and PAN, B. S. 1991. "Role of Collagen and Contractile Elements in Ultimate Tensile Strength of Squid Mantle." *J. Agric. Food Chem.* 39:1149–1154.

LOVE, M. 1970. *The Chemical Biology of Fishes.* London: Academic Press.

LOVE, M. 1980. *The Chemical Biology of Fishes.* Vol. 2. London: Academic Press.

LOVE, R. M., LAVETY, J., and GARCIA, N. G. 1972. "The Connective Tissues of Fish. VI. Mechanical Studies on Isolated Myocommata." *J. Food Technol.* 7:291–301.

LAVETY, J., AFOLABI, O. A., and LOVE, R. M. 1988. "The Connective Tissues of Fish. IX. Gaping in Farmed Species." *Internat. J. Food Sci. Technol.* 23:23–30.

MONTERO, P. 1988. *Caracterizacion y Estudio de Propiedades Functionales del Tejido Conectivo Muscular y Dermico de Merluza y Trucha. Efecto del Tratiamento.* Ph.D. thesis, Madrid.

MONTERO, P., and BORDERIAS, J. 1989a. "Changes in Hake Muscle Collagen During Frozen Storage Due to Seasonal Effects." *Internat. J. Refrig.* 12:220–223.

MONTERO, P., and BORDERIAS, J. 1989b. "Distribution and Hardness of Muscle Connective Tissue in Hake (*Merluccius merluccius L*) and trout (*Salmo irideus* (Gibb)." *Z. Lebensm. Unters. Forsch.* 189:530–533.

MONTERO, P., and BORDERIAS, J. 1989c. "Behaviour of Myofibrillar Proteins and Collagen in Hake (*Merluccius merluccius* L.) Muscle During Frozen Storage and Its Effect on Texture." *Z. Lebensm. Unters. Forsch.* 190:112–117.

MONTERO, P., and BORDERIAS, J. 1990a. "Effect of Rigor Mortis and Ageing on Collagen in Trout (*Salmo irideus*) Muscle." *J. Sci. Food Agric.* 52:141–146.

MONTERO, P., and BORDERIAS, J. 1990b. "Influence of Age on Muscle Connective Tissue in Trout (*Salmo irideus*)." *J. Sci. Food Agric.* 51:261–269.

MONTERO, P., and BORDERIAS, J. 1990c. "Gelification of Collagenous Material from Muscle and Skin of Hake (*Merluccius merluccius* L.) and Trout (*Salmo irideus* Gibb) According to Variation in pH and the Presence of NaCl in the Medium." *Z. Lebensm. Unters. Forsch.* 191:11–15.

MONTERO, P., and BORDERIAS, J. 1991. "Emulsifying Capacity of Collagenous Material from the Muscle and Skin of Hake (*Merluccius merluccius* L) and Trout (*Salmo irideus* (Gibb): Effect of pH and NaCl Concentration." *Food Chem.* 41:251–267.

MONTERO, P., BORDERIAS, J., TURNAY, J., and LEYZARBE, M. A. 1990. "Characterization of Hake (*Merluccius merluccius* L.) and Trout (*Salmo irideus* Gibb) Collagen." *J. Agric. Food Chem.* 38:604–609.

MONTERO, P., JIMENEZ-COLMENERO, F., and BORDERIAS, J. 1991. "Effect of pH and the Presence of NaCl on Some Hydration Properties of Collagenous Material from Trout (*Salmo irideus*) Muscle and Skin." *J. Sci. Food Agric.* 54:137–146.

OCHIAI, Y., KARIYA, Y., WATABE, S., and HASHIMOTO, K. 1985. "Heat-induced Tendering of Turban Shell (*Batillus cornutus*) Muscle." *J. Food Sci.* 50:981–984.

OTWELL, W. S., and GIDDINGS, G. G. 1980. "Scanning Electron Microscopy of Squid, Raw, Cooked and Frozen Mantle." *Marine Fish. Rev.* 42:67–73.

PEARSON, A. M., DUTSON, THAYNE R., and BAILEY, ALLEN J. (EDS.). 1987. "Collagen as a Food." In *Advances in Meat Research*. Vol. 4. New York: Van Nostrand Reinhold.

RAMSHAW, J. A. M., WERKMEISTER, J. A., and BREMNER, H. A. 1988. "Characterization of Type I Collagen from the Skin of Blue Grenadier (*Macruronus novaezelandiae*)." *Arch. Biochem. Biophys.* 267:497–502.

SADOWSKA, M., and SIKORSKI, Z. E. 1987. "Collagen in the Tissues of Squid *Illex argentinus* and *Loligo patagonica*—Contents and Solubility." *J. Food Biochem.* 11:109–120.

SATO, K., YOSHINAKA, R., SATO, M., and IKEDA, S. 1986a. "A Simplified Method for Determining Collagen in Fish Muscle." *Bull. Japan. Soc. Sci. Fish.* 52:889–893.

SATO, K., YOSHINAKA, R., SATO, M. and SHIMIZU, Y. 1986b. "Collagen Content in the Muscle of Fishes in Association with Their Swimming Movement and Meat Texture." *Bull. Japan Soc. Sci. Fish.* 52:1595–1600.

SATO, K., OHASHI, C., OHTSUKI, K., and KAWABATA, M. 1991. "Type V Collagen in Trout (*Salmo gairdneri*) and its Solublity Change during Chilled Storage of Muscle." *J. Agric. Food Chem.* 39:1222–1225.

SIKORSKI, Z. E., SCOTT, D. N., and BUISSON, D. H. 1984. "The Role of Collagen in the Quality and Processing of Fish." *Crit. Rev. Food Sci. Nutrition* 20:301–343.

TAKAHASHI, T., and YOKOYAMA, W. 1954. "Physicochemical Studies on the Skin and Leather of Marine Animals. XII. The Content of Hydroxyproline in a Collagen of Different Fish Skins." *Bull. Japan. Soc. Sci. Fish.* 20:525–529.

TAKEMA, Y., and KIMURA, S. 1982. "Two Genetically Distinct Molecular Species of Octopus Muscle Collagen." *Biochimica et Biophysica Acta.* 706:123–128.

YAMAGUCHI, K., LAVETY, J., and LOVE, R. M. 1976. "The Connective Tissues of Fish. VIII. Comparative Studies on Hake, Cod, and Catfish Collagens." *J. Food Technol.* 11:389–399.

YOSHINAKA, R., SATO, K., ANBE, H., SATO, M., and SHIMIZU, Y. 1988. "Distribution of Collagen in Body Muscle of Fishes with Different Swimming Modes." *Comp. Biochem. Physiol.* 89B(1):147–151.

ZHU, X., and KIMURA, S. 1991. "Thermal Stability and Subunit Composition of Muscle and Skin type I Collagen from Skipjack." *Nippon Suisan Gakkaishi.* 57:755–760.

6

The Involvement of Proteins and Nonprotein Nitrogen in Postmortem Changes in Seafoods

Zdzisław E. Sikorski and Bonnie Sun Pan

THE SEQUENCE OF CHANGES AND THE EFFECT OF HANDLING ON BOARD

Many nitrogenous compounds are involved in postmortem reactions in the muscles of animals, primarily as enzymes catalyzing the catabolysis of different tissue components. Thus, they contribute to changes in color, flavor, and rheological properties of the products. Furthermore, proteins and NPN participate also as substrates in reactions leading to the formation of products affecting the pH in the fish tissues. On the other hand, the rate and extent of changes in proteins are influenced by the products of degradation of saccharides and organic phosphates, as well as by the concentration of free ions, whereas these depend on the antemortem condition of the fish and on the temperature in the tissues after catch. A positive relationship of the enthalpy of activation of fish myofibrillar ATPase and the habitat temperature was demonstrated, from large negative values for cold-adapted species to high positive values for tropical species (Johnstone and Goldspink, 1975).

The typical sequence of events presented in Table 6–1, with the different stages well separated in time, applies only to rested fish killed

Table 6–1 Sequence of changes in nitrogenous compounds in caught fish

Stage After Catch	Principal Biochemical Changes	Sensory Effects
Prerigor	Enzyme-catalyzed oxidation of skin pigments; oxidation of blood and muscle chromoproteins due to anoxial conditions; possible denaturation of some proteins in mishandled tuna; hydrolysis of urea; deamination of adenosine phosphate and adenosine	Very slow loss of color and shine of the skin; slight darkening of the gills; possible discoloration and softening of tuna meat; development of ammoniacal odor in rays and sharks
Rigor mortis	Further changes in pigments; changes in binding of ions; interactions of contractile proteins; release of lysosomal enzymes; early proteolytic reactions	Onset, duration, and resolution of stiffness; further loss of shine and color of skin; darkening of the gills; slight loss in the typical fresh aroma
Loss of freshness	Further stages of proteolysis due to the endogenous muscle enzymes with formation of peptides and amino acids; advanced changes in pigments; decomposition of TMAO; development of volatile bases from various substrates	Further deterioration of color of the skin; browning of the gills; slight change in the color of the meat; loss in elasticity of the meat; development of fishy odor
Bacterial development	Advanced proteolysis catalyzed by endogenous and bacterial enzymes; formation of volatile compounds by decomposition of amino acids and other substrates	Slight bacterial slime on the surface; further discoloration of meat; development of plasticity of the meat; slight off-odor of the meat
Spoilage	Advanced stages of proteolysis; formation of putrid compounds and biogenic amines.	Thick bacterial slime on the surface; yellow discoloration of white fish meat; viscoplastic texture of belly flaps; heavy off-odor

rapidly and kept under refrigeration. In mishandled catch, at high ambient temperatures, especially in heavily feeding small fish like the Caspian sprats or capelin and in Antarctic krill, the rate of changes may be so high that the stages can hardly be distinguished. In such fish, rapid bacterial growth is possible within a few hours, resulting in the loss of nitrogenous compounds and massive accumulation of volatile degradation products. Therefore, efficient chilling, especially of large fish and of very big catches, is of prime significance for the quality of landed seafoods (Sikorski and Pan, 1991).

THE LOSS OF PRIME FRESHNESS

The Indices of Prime Freshness

The state of freshness is a very important factor contributing to the quality of any given species of seafood. In the most valuable species, this is reflected in the high price paid for the fish of prime freshness, suitable for being eaten raw. Seafoods in good condition and without even the slightest signs of spoilage may cost only about 10% of the price of fish of the same species just after catch.

The indices of prime freshness of fish are vivid colors of the skin, shiny surface, light-red gills, elastic body, and a delicate, desirable seaweedy flavor. The color of the surface is the most important of all freshness attributes because the price of many seafoods is directly related to the hue intensity or the hue itself.

The Color

The color of the skin, shell, and exoskeleton of aquatic animals is affected mainly by the carotenoid pigments. These polyenoic compounds are not very stable and can change or lose their color due to handling on board. In the skin of different species of fish lipoxygenase-type enzymes have been found to be capable of degrading astaxanthin, tunaxanthin, and β-carotene to colorless compounds (Simpson, 1982). In fish and crustaceans, the carotenoids are associated with proteins either as carotenoproteins or as lipoglycoproteins. The pigments in complexes with proteins are more stable than the free form. Dissociation of the protein moiety brings about changes in color. In direct sunlight, the stability of carotenoproteins is low and the color may decrease by 50% in a few hours.

The light-red color of the gills in fresh fish stems from the presence of oxygenated hemoglobin in the blood. As anoxial conditions in the body of fish set in gradually after catch, the color of the blood turns purple-red due to the reaction:

$$oxyhemoglobin \rightarrow hemoglobin$$

In large tuna, very rapid hauling the fish on board just after hooking may cause acidosis in the muscles due to strenuous struggling under conditions of a very restricted supply of oxygen and a high rate of anaerobic metabolism. If the handling on board does not provide for rapid chilling of the fish, the myoglobin may be oxidized to metmyoglobin and some muscle proteins may be denatured at the low pH and high body temperature. Furthermore, under such conditions of fishing, there is, in the tissues, a favorable environment for high activity of proteolytic enzymes, especially of the calcium-activated neutral proteinases (CANP). As a result, the "burnt tuna" effect can be seen, that is, the meat is soft and pale muddy brown or turbid, has a stringent aftertaste, and is not suitable for raw consumption (Davie and Sparksman, 1986). The hypothesis on the etiology of burnt tuna, presented by Watson et al. (1988), is based on postmortem activation of CANP and on enhancement of the action of these enzymes by high levels of neurotransmitters-hormones norepinephrine and epinephrine (catecholamines). According to this hypothesis, the fighting of the fish on hook under conditions impairing oxygen supply leads to a rapid increase of Ca^{2+} in the sarcoplasm due to depletion of ATP reserves. This brings about an activation of CANP, attacking troponin, tropomyosin, sarcoplasmic reticulum, and mitochondria, and leads to an increased rate of disintegration of the Z-disks in the myofibrils. Furthermore, struggling increases the level of blood catecholamines which enhance the action of CANP. The catecholamines can be degraded in the gills of fish. Therefore, in longline-caught tuna, killed several hours after hooking, their concentration as well as the incidence of burnt tuna is much lower than in handline-caught fish.

The Flavor

The desirable fresh aroma of seafoods has been associated with the presence of a number of 6-, 8-, and 9-carbon aldehydes, ketones, and alcohols generated from polyunsaturated fatty acids due to the activity of various lipoxygenases (Lindsay, 1991b). Among the volatiles in fresh whitefish, Josephson et al. (1983) identified 12 such compounds present in concentrations above their reported threshold. Several of these components were identified also in fresh Atlantic and Pacific oysters (Josephson et al., 1985). During storage of fish, the original fresh aroma changes rapidly as the result of biochemical, bacteriological, and chemical processes. Microbial conversion of some carbonyl compounds to alcohols has been found to be responsible for the loss of the initial flavor of seafoods (Lindsay, 1991a). The rate of these changes depends mainly on the temperature of storage after catch, although the composition of the atmosphere in the package, as well as irradiation, affect the pathways of the processes. In the later stages of storage, volatile nitrogenous and sulphur compounds are responsible for putrid odor.

In several species of rays and sharks, the endogenous muscle urease may cause in the fresh fish, early after catch, a detectable odor of ammonia due to hydrolysis of urea. Such meat may have a bitter taste. The sensory recognition threshold of urea in water solution is about 0.5% and is somewhat higher in fresh meat. Special treatments of various species of sharks during handling and in culinary preparation are necessary to make the fish suitable for human consumption. Early and complete bleeding by severing the tail vein, gutting, filleting, and washing removes much of the urea. Further reduction of the urea content in the muscles, to levels below 1.2%, can be achieved by leaching about 2 h at room temperature in a solution, 1 : 5, containing 2–2.5% NaCl and 0.02–0.04% acetic acid. The meat of the great blue shark, cooked after such treatment, has very acceptable sensory properties (Koczot et al., 1981).

In shrimp muscles, ammonia can be generated soon after catch by deamination of adenosine and AMP by endogenous enzymes. These enzymes, although very active at room temperature, have only a marginal activity in iced shrimp. Rapid chilling can prevent the accumulation of ammonia during the first days after catch (Finne, 1982).

In the early postmortem flavor changes in sea fish, a crucial role is played by the enzymatic breakdown of nucleotides. The nonvolatile components produced from the degradation of nucleotides can either enhance the desirable flavor, as IMP which at a certain concentration gives the desirable full flavor to the fish (Murata and Sakaguchi, 1989), or may add undesirable notes, as hypoxanthine contributing to bitterness (Lindsay, 1991). The rate of the reactions responsible for the loss of prime freshness in fish is controlled by the temperature of storage (Sikorski, 1990) but is also affected by the habitat temperature of the fishes (Tsuchimoto et al., 1986).

Other Properties

The death conditions of fish were found to affect not only the rate of freshness degradation but also the functional properties of fish proteins. Mackerel which struggled before killing had poorer gel-forming ability and was more susceptible to the modori degradation of the gel than the instantaneously killed fish (Shimizu and Kaguri, 1986).

THE INVOLVEMENT OF PROTEINS AND NONPROTEIN NITROGEN IN RIGOR MORTIS

Rigor mortis is a biophysical phenomenon related to the state of the myofibrillar proteins. It is caused by biochemical changes in the muscles involving ATP degradation, catalyzed by endogenous enzymes.

Most of the published data on rigor mortis in fish regard species of cold and temperate waters. In cold-water species, rapid chilling of the fish in ice generally increases the time of onset and the duration of rigor, thus extending the state of prime freshness. In some tropical fish, however, rapid chilling may induce almost immediate cold-shock stiffening that is not accompanied by muscle contraction nor by a correspondingly fast depletion of ATP (Curran et al., 1986). In beef muscle, the dephosphorylation of ATP and the onset of rigor mortis is affected by temperature in a way that cannot be described by the van't Hoff relationship. At temperatures of about 0°C, the rate of these processes is significantly higher than at 15–20°C and the rapid onset of rigor mortis is called cold shortening. In the muscles of plaice, the rate of ATP degradation and of onset of rigor at 0°C is higher than at 5–15°C, although the activity of the myofibrillar and sarcoplasmic ATP-ases and of the creatine kinase are lower than at 10°C (Iwamoto et al., 1987, 1988). Also, in sardine and mackerel, rigor mortis at 10°C sets in later than at 0°C (Watabe et al., 1989).

Acclimation of eurythermal temperate zone fish, for example, goldfish and carp, leads to compensatory changes in myofibrillar ATPase activity, thus affecting the progress of rigor mortis. In carp living in water at 16–17°C, acclimation at 10°C or 30°C caused rigor mortis to appear after 32 h and 16 h after death, respectively (Watabe et al., 1990).

Under the same temperature conditions, the time of onset of rigor after death and the duration of stiffness varies with the fish species. In iced tilapia, maximal rigor was observed 9 h after death and full rigor lasted 4 days (Toyohara and Shimizu, 1988), whereas in grey mullet it appeared after 2 and 15 h, respectively (Yeh and Pan, 1992).

Treatment of the fish on board may affect the time to onset of rigor. Spiking of the troll-caught fish kahawai (*Arripis trutta*) followed by chilling delayed both the onset of rigor by 1.2 h and the decrease in pH in the meat during the early stages of rigor by slowing-down the degradation of ATP and production of lactate (Boyd et al., 1984).

AUTOLYSIS

The autolytic processes in fish are manifested primarily by changes in the rheological properties and by marked alteration of the odor of the muscles. The viscoelastic fresh meat, with the predominant elastic element, turns gradually into a viscoplastic body due to proteolytic degradation of some of the structural proteins.

The texture of unprocessed sea fish may be an important quality criterion for species which are consumed raw in the form of sushi or sushimi. There is a general view that fish, just as the meat of slaughter animals, is tough when in rigor. However, in the muscles of fish belonging to several species, the first signs of tenderization were found early after catch, still before resolution of rigor mortis (Ando et al., 1991). The initial change in the characteristic rheological properties of fresh fish is catalyzed by endogenous muscle proteinases, as the muscles of fish just after capture are not significantly contaminated by bacteria. In the later stages the enzymes of the digestive tract, as well as bacterial proteinases, participate also in the process. An important role is played by the proteinases from the digestive tract which are induced by feed ingest. Therefore, autolytic changes appear early after catch, primarily in small, feeding fish.

The involvement of endogenous muscle proteinases has been extensively studied in meat with respect to beef aging; thus, the effect of cathepsins and of other muscle proteinases has been investigated. In evaluating the possible participation of these enzymes in the autolytic processes in fish, the following factors must be considered:

- The presence and seasonal change in concentration of the enzymes in the muscles of seafoods
- The susceptibility of different muscle structural proteins in situ to degradation by the respective enzymes
- The synergistic effect in proteolytic changes catalyzed by several different enzymes
- The activity of the enzymes in the pH range 5.4–7.0 toward structural muscle proteins
- The effect of temperature on the enzyme activity, with special emphasis of chill storage conditions

- The required concentration of the activating factors, for example, Ca^{2+}
- The effect of endogenous protein inhibitors on the activity of the respective enzymes in situ

In evaluating the effect of pH on the role of various enzymes in autolysis, it should be considered that the activity of different proteinases may be as high as over 50% of the maximum value at pH values even 1 or 2 units away from the pH optimum. Thus, not only the activity at optimum pH, but also in a broader pH range, should be considered. The same applies also to the effect of temperature.

SPOILAGE CHANGES

The advanced stages of quality deterioration of unfrozen seafoods are characterized by changes in nitrogenous and sulphur compounds caused principally by the spoilage microflora (Herbert et al., 1971). The odor of spoiling fish results from bacterial decomposition of the products of protein hydrolysis and other NPN components of the tissues. Besides TMA, DMA, methylamine, and ammonia formed from TMAO, urea, lecithins, amino acids, and nucleotides, indole, and skatole, as well as volatile sulfur-containing compounds, are generated: mainly hydrogen sulfide, methanethiol, ethanethiol, aminoethanethiol, thiopropionic acid, and dimethyl sulfide. Hydrogen sulfide may be produced in fish meat at a high rate due also to catalysis of the enzymes present in fish entrails (Hanusardottir, 1983). Accumulation of these volatiles in the tissues makes the seafood unfit for human consumption because of the repelling odor (Table 6–2).

Stale fish may contain several NPN compounds which may be harmful to consumers, mainly biogenic amines and some bacterial toxins. Decarboxylation of amino acids, especially in scombroid fish and mahi mahi, leads to the formation of diamines:

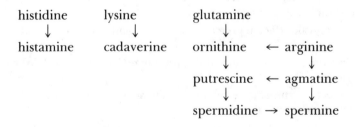

Table 6–2   Odor detection threshold of volatile compounds typical of spoiling seafoods

Compound	Threshold[a] (ng/dm^3 air)	Compound	Threshold[b] (μg/dm^3 Water)
Skatole	0.04×10^{-2}	Methional	0.2–5.0
Ethanethiol	0.04	Ethanethiol	1–10
Propanethiol	0.23	Propanethiol	3.1
Hydrogen sulfide	0.92	Hydrogen sulfide	0.5–10
Methanethiol	2.2	Methanethiol	0.02–2.1
Ammonia	26	Buthanethiol	1.1
Dimethylsulfide	51	Dimethylsulfide	0.3–12
Dimethylamine	1100	Diethylsulfide	4.8
Trimethylamine	9600	Dimethyldisulfide	2.1–7.6

Sources: (a) Summer (1963); (b) Baryłko-Pikielna et al. (1976).

The concentration of these amines in the fish flesh increases with the time of storage. Although histidine decarboxylase has been found in the muscles and liver of many fish species, this enzyme is generally hardly responsible for the generation of histamine. However, some cases of high contents of histamine, up to about 5.5 mg/g in the meat of fresh tuna are known (Yamanaka et al., 1982). Principally, the biogenic amines are generated as a result of the metabolism of several species of bacteria, mainly *Proteus morganii* and *Enterobacter aerogenes* (Eitenmiller et al., 1982; Frank, 1985; Sikorski et al., 1990). Although the optimum temperature for bacterial decarboxylation of histidine in fishery products is in the range of 20–40°C, there are also bacteria capable of accumulating large amounts of histamine even at temperatures as low as 5°C (Okuzumi et al., 1981).

Bacterial toxins are formed in fishery products mainly in advanced stages of spoilage. The most severe hazards are associated with the possible presence of the toxins of *Clostridium botulinum* (Hobbs, 1976) and *Staphylococcus aureus*. The *Cl. botulinum* toxins are relatively sensitive to thermal denaturation. The enterotoxins of *Staphylococcus aureus* are very resistant to proteolysis by trypsin, chymotrypsin, rennin, and papain, as well as to changes in pH in the range 3–10 and to heating.

USE OF PROTEIN AND NONPROTEIN-NITROGEN METABOLITES AS FRESHNESS INDICATORS

As the concentration of NPN and sulphur-containing compounds in the muscles after catch changes due to the activity of endogenous and bacterial enzymes, it can serve as a measure of freshness or staleness of sea-foods (Sikorski et al., 1990). Much attention has been paid to the accumulation of volatile bases, especially of TMA in the muscles of seafoods during storage. Their initial contents and storage changes in fish and shellfish, however, are highly affected by the species and by the catching season. Especially high contents and seasonal fluctuations of volatile bases have been found in Antarctic krill (Kołakowski and Moreno, 1987). Thus, these compounds cannot be applied as reliable freshness indicators regardless of the species of fish. Recently, the contents of H_2S in gill air of cod has been found correlated with the time of storage of the fish on ice, with a 95% prediction interval of ± 3 days. No measurements, however, could be made until day 6 in ice (Strachan and Nicholson, 1992).

Generally, the compounds which are degraded or generated in reactions catalyzed by endogenous enzymes are rather good indices of the initial freshness changes, whereas a high concentration of bacterial metabolites is evidence of spoilage. Initially, the determination of hypoxanthine was regarded as most suitable for evaluating the freshness of fish in the early days after catch. However, as in many species, inosine is accumulated instead of hypoxanthine (Echira and Uchiyama, 1973), the K-value, that is, the percent ratio of inosine plus hypoxanthine to the sum of ATP and its degradation products, has been found to be a more useful indicator.

The quality of abalone as food is highly affected by the stress during handling and transporting live. Recently, tauropine, a product of condensation of pyruvate and taurine, has been shown to be a suitable indicator of the metabolic stress in these animals (Baldwin et al., 1992).

REFERENCES

ANDO, M., TOYOHARA, H., SHIMIZU, Y., and SAKAGUCHU, M. 1991. "Postmortem Tenderization of Fish Muscle Proceeds Independently of Resolution of Rigor Mortis." *Nippon Suisan Gakkaishi.* 57:1165–1169.

BALDWIN, J., WELLS, R. M. G., LOW, M., and RYDER, J. M. 1992. "Tauropine and D-Lactate as Metabolic Stress Indicators During Transport and Storage of Live Paua (New Zealand Abalone) (*Haliotis iris*)." *J. Food Sci.* 57:280–282.

BARYŁKO-PIKIELNA, N., MIELNICZUK, Z., and DANIEWSKI, M. 1976. "Lotne Związki Siarkowe w Mięsie oraz ich Prekursory (Przegląd Aktualnego Stanu Badań)." *Rocz. Inst. Przem. Mies. Tłuszcz.* 13:67–78.

BOYD, N. S., WILSON, N. D., JERRETT, A. R., and HALL, B. I. 1984. "Effects of Brain Destruction on Post Harvest Muscle Metabolism in the Fish Kahawai (*Arripis trutta*)." *J. Food Sci.* 49:177–179.

CURRAN, C. A., POULTER, R. G., BRUETON, A., and JONES, N. S. D. 1986. "Cold Shock Reactions in Iced Tropical Fish." *J. Food Technol.* 21:289–299.

DAVIE, P. S., and SPARKSMAN, R. I. 1986. "Burnt Tuna: An Ultrastructural Study of Postmortem Changes in Muscle of Yellowfin Tuna (*Thunnus albacores*) Caught on Rod and Reel and Southern Bluefin Tuna (*Thunnus maccoyi*) Caught on handline or longline." *J. Food Sci.* 51:1127–1128, 1168.

ECHIRA, S., and UCHIYAMA, H. 1973. "Formation of Inosine and Hypoxanthine in Fish Muscle During Ice Storage." *Bull. Tokai Reg. Fish. Res. Lab.* 75:63–73.

EITENMILLER, R. R., ORR, J. H., and WALLIS, W. W. 1982. "Histamine formation in Fish: Microbiological and Biochemical Conditions." Pp. 39–50. In *Chemistry and Biochemistry of Marine Food Products,* edited by R. E. Martin, G. E. Flick, C. E. Hebard, and D. W. Ward. New York: Van Nostrand Reinhold.

FINNE, G. 1982. "Enzymatic Ammonia Production in Penaied Shrimp Held on Ice." Pp. 323–331. In *Chemistry and Biochemistry of Marine Food Products,* edited by R. E. Martin, G. E. Flick, C. E. Hebard, and D. W. Ward. New York: Van Nostrand Reinhold.

FRANK, H. A. 1985. "Histamine Forming Bacteria in Tuna and Other Marine Fish." Pp. 2–3. In *Histamine in Marine Products: Production by Bacteria, Measurement and Prediction of Formation,* edited by Bonnie Sun Pan and David James. FAO Fish Technol. Paper 252. Rome: Food and Agriculture Organization.

HANUSARDOTTIR, M. 1983. "Some Aspects Concerning the Causal Relation in Connection with the Development of Hydrogen sulfide (H_2S) in Codfish." *Nor. Vet.-Med.* 35:300–305.

HERBERT, R. A., HENDRIE, M. S., GIBSON, D. M., and SHEWAN, J. M. 1971. "Bacteria Active in the Spoilage of Certain Sea Foods." *J. Appl. Bact.* 34:41–50.

HOBBS, G. 1976. "*Clostridium botulinum* and Its Importance in Fishery Products." Pp. 135–185. In *Advances in Food Research,* edited by C. O. Chichester, E. M. Mrak, and G. F. Stewart. Vol. 22. New York: Academic Press.

IWAMOTO, M., YAMANAKA, H., WATABE, S. D., and HASHIMOTO, K. 1987. "Effect of Storage Temperature on Rigor Mortis and ATP and Creatine Phosphate Degradation in Plaice *Paralichthys olivaceus* Muscle." *J. Food Sci.* 52:1514–1517.

IWAMOTO, M., YAMANAKA, H., ABE, H., USHIO, H., WATABE, S., HASHIMOTO, K. 1988. ATP and creative phosphate breakdown in spiked plaice muscle during storage, and activities of some enzymes involved. *J. Food Sci.* 53:1662–1665.

JOHNSTONE, I. A., and GOLDSPINK, G. 1975. "Thermodynamic Activation Parameters of Fish Myofibrillar ATPase Enzyme and Evolutionary Adaptations to Temperature." *Nature* 257:620–622.

JOSEPHSON, D. B., LINDSAY, R. C., and STUIBER, D. A. 1983. "Identification of Compounds Characterizing the Aroma of Fresh Whitefish (*Coregonus clupeaformis*)." *J. Agric. Food Chem.* 31:326–330.

JOSEPHSON, D. B., LINDSAY, R. C., and STUIBER, D. A. 1985. "Volatile Compounds Characterizing the Aroma of Fresh Atlantic and Pacific Oysters." *J. Food Sci.* 50:5–9.

KOCZOT, A., SZEWCZUK, M., ZALEWSKI, J., ŁĄDKOWSKA, M., BARSKA, I., BRZESKI, M., KORZEKWA, W., and PRĘDA, M. 1981. "Przydatność Technologiczna Ryb Chrzęstnoszkieletowych i Kostnoszkieletowych Otwartego Oceanu." Research Report No. T–133. Sea Fisheries Institute, Gdynia.

KOŁAKOWSKI, E., and MORENO, M. M. 1987. "Zawartość Lotnych Zasad w Świeżym Krylu Antarktycznym (*Euphausia superba Dana*)." *Bromat. Chem. Toksykol.* 20:36–41.

LINDSAY, R. C. 1991a. "Chemical Basis of the Quality of Seafood Flavors and Aromas." *Marine Technol. Soc. J.* 25:16–22.

LINDSAY, R. C. 1991b. "Flavor of Fish." Paper read at 8th World Congress of Food Science and Technology, 29 September–4 October, Toronto, Canada.

MURATA, M. and SAKAGUCHI, M. 1989. "The Effects of Phosphatase Treatment of Yellowtail Muscle Extracts and Subsequent Addition of IMP on Flavor Intensity." *Nippon Suisan Gakkaishi.* 55:1599–1603.

OKUZUMI, M., OKUDA, S., and AWANO, M. 1981. "Isolation of Psychrophilic and Halophilic Histamine-forming Bacteria from *Scomber japonicus*." *Bull. Japan. Soc. Sci. Fish.* 47:1591–1598.

PAN, B. S. and YEH, W. T. 1993. "Biochemical and morphological changes in grass shrimp (*Penaeus monodon*) muscle following freezing by air blast and liquid nitrogen methods. *Food Biochem.* in press.

SHIMIZU, Y., and KAGURI, A. 1986. "Influence of Death Condition and Freshness on the Gel-Forming Property of Fish." *Bull. Japan. Soc. Sci. Fish.* 52:1837–1841.

SIKORSKI, Z. E. 1990. "Chilling of Fresh Fish." Pp. 93–109. In *Seafood: Resources, Nutritional Composition and Preservation,* edited by Z. E. Sikorski. Boca Raton, FL: CRC Press.

SIKORSKI, Z. E., and PAN, B. S. 1991. "Preservation of Seafood Quality." Paper read at 8th World Congress of Food Science and Technology, 29 September–4 October, Toronto, Canada.

SIKORSKI, Z. E., KOŁAKOWSKA, A., and BURT, J. R. 1990. "Postharvest Biochemical and Microbial Changes." Pp. 55–75. In *Seafood: Resources, Nutritional Composition, and Preservation,* edited by Z. E. Sikorski. Boca Raton, FL: CRC Press.

SIMPSON, K. L. 1982. "Carotenoid Pigments in Seafood." Pp. 115–136. In *Chemistry and Biochemistry of Marine Food Products,* edited by R. E. Martin, G. E. Flick, C. E. Hebard, and D. W. Ward. New York: Van Nostrand Reinhold.

STRACHAN, N.J.C., and NICHOLSON, F. J. 1992. "Gill Air Analysis as an Indicator of Cod Freshness and Spoilage." *Internat. J. Food Sci. Technol.* 27:261–269.

SUMMER, W. 1963. *Methods of Air Deodorization.* Amsterdam: Elsevier.

TOYOHARA, H., and SHIMIZU, Y. 1988. "Relation of the *Rigor Mortis* of Fish Body and the Texture of the Muscle." *Nippon Suisan Gakkaishi.* 54:1795–1798.

TSUCHIMOTO, M., MISIMA, T., UTSUGI, T., KITAJIMA, S., YADA, S., SENTA, T., and YASUDA, M. 1986. "The Speed of Lowering in Freshness of Fishes in Several Waters and the Effect of the Habitat Temperature on the Speed." *Bull. Japan. Soc. Sci. Fish.* 52:1431–1441.

WATABE, S., USHIO, H., IWAMOTO, M., KAMAL, M., IOKA, H., and HASHIMOTO, K. 1989. "Rigor-Mortis Progress of Sardine and Mackerel in Association with ATP Degradation and Lactate Accumulation. *Nippon Suisan Gakkaishi.* 55:1833–1839.

WATABE, S., HWANG, G. C., USHIO, H., and HASHIMOTO, K. 1990. "Changes in Rigor Mortis Progress of Carp Induced by Temperature Acclimation." *Agric. Biol. Chem.* 54(1):219–221.

WATSON, C., BOURKE, R. E., and BRILL, R. W. 1988. "A Comprehensive Theory on the Etiology of Burnt Tuna." *Fishery Bull.* 86 (2): 367–372.

YAMANAKA, H., SHIOMI, K., KIKUCHI, T., and OKUZUMI, M. 1982. "A Pungent Compound Produced in the Meat of Frozen Yellowfin Tuna and Marlin" *Bull. Japan. Soc. Sci. Fish.* 48:685–689.

7

THE EFFECT OF HEAT-INDUCED CHANGES IN NITROGENOUS CONSTITUENTS ON THE PROPERTIES OF SEAFOODS

Zdzisław E. Sikorski and Bonnie Sun Pan

INTRODUCTION

Heating, as applied in seafood technology, affects the rate of enzymatic processes, brings about denaturation and subsequent interactions of proteins, and accelerates chemical reactions of different tissue components. Therefore, heated seafoods change in color and in rheological properties, lose part of their water-retention capacity, and develop new, mostly very desirable flavor characteristics. The character and extent of these changes are affected by the rate, duration, and temperature of heating, as well as by the composition, biochemical state, and integrity of the food structure, the pH of the environment, the presence of different added substances, mainly salts, macromolecular extenders, and polyols, and the access of air (Aitken and Connell, 1979; Pan, 1990). Investigations on the effect of these factors should provide the fish processor with knowledge necessary to control the quality of cooked seafood products. Proper choice of heating rate, duration, temperature, and humidity may prevent losses in hot smoking of fish (Sikorski et al., 1984). Applying high enough time/temperature in cooking of crab may minimize leaching losses of soluble proteins and discoloration as well as facilitates the picking of the meat (Zaitsev et al., 1969; Dowdie and Biede, 1983).

ENZYMATIC CHANGES IN HEATED SEAFOOD PRODUCTS

In fish processing, contrary to what is purposely carried out in many other food industries, the objective of heating with regard to enzymes is generally to rapidly inactivate the endogenous and bacterial enzymes to avoid deterioration in the quality of the products. Generally, cooking easily destroys the enzyme activity in fish, for example, the thiaminase which can induce severe loss of thiamine in animal feeds containing raw fish offal. Cod muscle catalase can be totally inactivated by heating 10 min at 50°C, whereas bacterial catalase remains unaltered, which makes it possible to apply the catalase test for a quick estimation of bacterial contamination of the fillets (Boismenu et al., 1990).

The conditions of heating during some stages of fish processing, however, may also favor undesirable enzymatic changes. One such process taking place during heating of fish meat is the modori phenomenon, that is, weakening of the structure of fish gel at a temperature of about 60°C, caused by heat-stable alkaline muscle proteinase. This enzyme has been found in carp, Atlantic croaker, white croaker, Atlantic menhaden, and mullet (Hamann et al., 1990). The gel degradation is caused by a breakdown of the myosin heavy chain and by the accumulation of small peptides. The optimum temperature of the alkaline muscle proteinase of carp is 65°C and that of mackerel is 60°C. The pH optimum is 8. The alkaline proteinase may also contribute to honeycombing because mackerel fillets, injected with this enzyme and cooked in conditions corresponding to those applied in precooking in the canning industry, showed indentations of myomeres (Pan et al., 1986).

Recently, other latent heat-activated muscle proteinases were discovered. They also participate in the weakening of fish gels and in degradation of the heavy chain of myosin due to heating to about 60°C. One such proteinase was found in the sarcoplasm in threadfin-beam and rainbow trout, whereas an other, tightly associated with myofibrils, was found in oval-filefish and crucian carp (Toyohara et al., 1990). In tilapia muscles, both types of enzymes are present (Toyohara et al., 1990a). The activity of the myofibril-associated enzyme in situ can be decreased by the addition of the sarcoplasmic fraction, which probably contains some thermolabile inhibitors. Both enzymes can be inhibited by certain factors present in fish blood because bleeding was found to promote the modori effect (Toyohara et al., 1990b). The proteinases tightly associated with myofibrils in crucial carp were differentiated into a species activated at 50°C, with a pH optimum at 6.6–8.0, and one activated at 60°C with a pH optimum at 6.2–7.1 (Kinoshita et al., 1990).

CHANGES IN RHEOLOGICAL PROPERTIES AND HYDRATION OF SEAFOODS

The Role of Different Groups of Proteins

The rheological properties of heated seafoods are mainly affected by the state of the histological structure and by the ability of muscle proteins to retain their native molecular properties, hydration, and gel-forming capacity. Among the three main groups of muscle proteins, the sarcoplasmic proteins have no significant impact on the functional properties. They do not contribute to the histological structure of the tissues and easily lose solubility when heated. However, some of the glycolytic enzymes, particularly aldolase and glyceraldehyde–3-phosphate dehydrogenase, may affect the rheological properties of fish gels. They bind strongly to F-actin and actomyosin and may, thus, remain in surimi after the washing process. During cooking, these heat-coagulable proteins impair the gel-forming ability of actomyosin (Nakagawa and Nagayama, 1989).

The thermal changes in the myofibrillar proteins result in increased toughness of the fish fillet, whereas heat-induced transformations of collagen are responsible for making the meat more tender, for flaking of the fillets, and for splitting of the fish skin. Collagen changes may also bring about losses in hot smoking due to breaking of the tissues above the point of suspension of the fish on hooks or spits (Sikorski et al., 1984).

The thermal changes in different proteins are affected by their mutual interactions in the tissues or minces, as well as by the pH and the presence of salts (Beas et al., 1990). The effect of heating on fish proteins has been treated also in Chapters 4, 5, and 10.

A typical change in the rheological properties of fish meat due to heating (Fig. 7–1) is a reflection of the thermal behavior of myofibrillar proteins and collagen. Denaturation of both groups of proteins in the temperature range 30–70°C brings about a significant shrinkage of the meat, with a simultaneous loss of fluid. This stage corresponds to the increase in toughness presented in Fig. 7–1. Steaming of fish as applied in canneries may cause a loss of up to about 20% of fluid containing 1–3% crude protein. The cooking drip in krill is about 30% and, in extreme

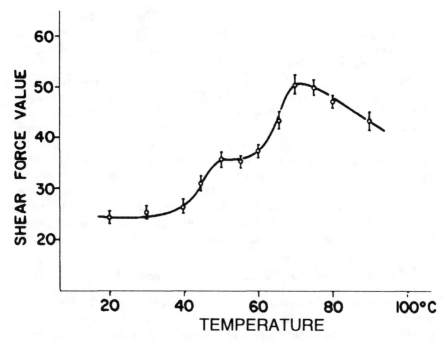

Figure 7–1   A typical relationship between the cooking temperature and the shear force values for postrigor pacific cod muscles.

Source: Dunajski (1979).

cases, may reach 60%. The release of water from cooked herring increases with the temperature of heating although at 60°C, there is a minimum on the curve of water loss versus temperature (Beraquet et al., 1984). Further heating causes rupturing of the bonds which buttress the structure of the collagen fibers and fibrils, uncoiling of the superhelix, and a transition into gelatin, with corresponding loss in toughness of the meat. The ultrastructural changes due to heating were investigated in detail in the ordinary and dark muscles of yellowfin tuna by Kanoh et al. (1988) and in squid mantle by Otwell and Hamann (1979).

Some rheological changes in heated eggs, meat, and fish are brought about by aggregation of proteins due to the formation of new hydrogen and disulfide bonds, as well as hydrophobic interactions (Buttkus, 1974). In the temperature range used for cooking of foods, the thiol–disulfide exchange reactions may be mainly responsible for the aggregation of proteins. Furthermore, the SH groups in amino acid residues in proteins may also react with formaldehyde (FA). In fish meat, FA may be formed by the decomposition of TMAO, either catalyzed by the TMAO demethylase, which retains its activity up to 88°C (Lall 1975) or can be generated in reactions catalyzed by cations or other compounds present in the muscle tissues (Spinelli and Koury, 1979, 1981). The involvement of SH groups and FA in the rheological changes in cooked squid has been shown (Synowiecki and Sikorski, 1988). After 45 min at 98°C, the contents of SH groups in squid mantle decreased by about 30%, although the total contents of these groups after reduction by $NaBH_4$ did not change significantly. A slightly higher decrease in the contents of SH groups due to cooking, not accompanied by formation of H_2S, was found recently in seal muscle proteins by Synowiecki and Shahidi (1991). The cooked squid contained about 18 times more DMA and only twice as much free FA as the raw samples. About 95% of the generated FA was not available for direct determination, apparently mainly due to reactions with proteins, however, not involving SH groups.

The Effect of Species Characteristics and Treatment

The effect of heating to any given temperature on the hydration and texture of seafood products depends on the pH of the meat and on the fish species. At low pH, the toughness of cooked cod increases in the temperature range 40–70°C as shown in Fig. 7–1, but at pH 7.1 there is almost no change in texture of the muscle due to heating (Dunajski, 1977). The importance of the species characteristics and of the condition of the fish on the rheological properties of the cooked product is reflected in the traditionally recognized suitability of different species and stocks for canning, roasting, and other methods of culinary preparation.

A very mild heat treatment must be applied to peeled krill tail meat for making the product palatable (Table 7–1). Heating krill above 50°C brings about a significant deterioration of the texture and flavor of the product and increases the total thawing and cooking loss to about 60% (Bykowski and Kołodziejski, 1983).

Many species of squid used for food differ significantly in texture after cooking, the most tender being *Todarodes pacificus, Loligo vulgaris,* and *Loligo edulis* (Kreuzer, 1984). The rheological properties of cooked squid are also affected by the treatment of the raw material and by the conditions of cooking. According to Stanley and Hultin (1982), frozen storage brings about a slight increase in hardness of the mantle. The state of freshness prior to freezing has, apparently, an effect on the texture of squid *Illex argentinus* cooked after prolonged frozen storage. Squid with distinct discolorations of the skin and of the muscle layer just

Table 7–1 The effect of heating temperature on the properties of krill tail meat

Temperature (°C)	Drip Loss (%)	Texture of Meat
40	23.2	Elastic, juicy
50	26.1	Slightly tough and dry
60	28.3	Tough, brittle, and dry
70	32.7	Tough, brittle, and dry

Source: Bykowski and Kołodziejski (1983).

under the skin were found to be about 30% less tough than fresher, undiscolored mantle. Holding thawed squid prior to cooking 18 h at room temperature or 2 h at 35°C did not improve the texture of the cooked product, although it caused significant proteolysis of myosin. Also, curing of the raw muscle in 2% acetic acid solution caused detectable proteolytic changes, whereas the hardness of the cooked product increased significantly and the cooking loss was 70% larger than in the controls. On the other hand, soaking the mantle in 5% NaCl solution or treatment with polyphosphates decreased the cooking loss and toughness, having no effect on the proteolytic activity in squid meat (Sikorski and Kołodziejska, 1986).

Some published data indicate that significant texture changes in squid mantle take place only during the first 8 min of cooking (Stanley and Hultin, 1982). On the other hand, the hardness of frozen stored squid *Illex argentinus* decreases during at least 60 min boiling in 2% NaCl solution, although the weight loss does not continue after about 15 min (Sikorski and Kołodziejska, 1986).

Properties of Comminuted Products

The texture of comminuted cooked fish products is affected by the state of the proteins in the raw material, but the applied additives influencing the gelling, emulsifying, and water-holding properties of the mince, and by the conditions of processing (Pan, 1990). Protein denaturation in the frozen stored raw material has a very significant impact on the texture of the cooked product. This will be discussed extensively in Chapters 8 and 10.

In comminuted cooked sausages, produced from fresh and frozen stored raw fish without the addition of starch or other macromolecular extenders, the rheological attributes, which are highly correlated to the heptaesthetic data, can be represented by the yield value. The texture of sausages made of cod mince was found acceptable when the yield value was not lower than $3 \times 10^4 \mathrm{Nm}^{-2}$. At constant shear stress, a highly significant correlation exists between the shear rate in minces of fresh and frozen Baltic cod and the yield value of the product after a standard heat treatment. Thus, the texture of the sausages can be controlled by using proper proportions of the fresh and frozen stored fish, based on the viscosity measurement in the mince (Sadowska and Sikorski, 1977).

CHANGES IN COLOR

Heating to cooking, frying, and sterilizing temperature brings about significant changes in the color of seafoods due to denaturation and oxidation of proteins, as well as the formation of colored compounds with involvement of H_2S released from amino acids and in Maillard-type reactions.

The meat of white fish gradually turns in color to whitish-gray and opaque in the temperature range 35–60°C. Red meat, if not cured, generally changes to grayish-brown or brown as a result of denaturation and oxidation of the hemoproteins. In horse mackerel fillets, the color starts to change at about 35°C and is stabilized as grayish-pink at about 80°C.

In canned products, black deposits os sulfides may occur, as results of the reaction of hydrogen sulfide with ions. The black discolorations are especially frequent in canned bivalves due to the presence of iron and copper. The latter may be released from hemocyanin from clam blood by the heat treatment. The extent of blackening is affected by the amount of H_2S produced from cysteine, cystine, methionine, and taurine and may be also controlled by additives and factors suppressing the concentration of ionized H_2S (Boon, 1977). Blackening is especially intensive in crab meat of inferior freshness and is reduced at higher acidity of the medium.

Blue discolorations of various shades may occur in canned crabmeat containing more than the average amount of copper, usually above 2 mg/100 g. Generally, the blueing occurs on the surface of the meat and in the coagulated blood. In the reactions leading to this heat-induced discoloration, hemocyanin and H_2S are involved (Motohiro, 1982), but the state of freshness of the crab as well as the parameters of heating are important. Yellow discolorations may be found in canned crabmeat prepared from raw material affected by tyrosinase (Zaitsev et al., 1969). Babbitt (1982) presented a scheme for blueing of dungeness crabmeat, involving enzymatic oxidation of tyrosine to melanins prior to cooking, followed by nonenzymatic oxidation and polymerization of the phenolic derivatives, with the formation of colored compounds in the presence of metals.

Browning due to the Maillard-type reactions has a significant impact on the sensory quality only in white-meat fishery products. The extent of discoloration can be reduced by enzymatic decomposition or leaching of the saccharides mainly involved in these reactions.

DEVELOPMENT OF VOLATILE FLAVOR COMPOUNDS

Introduction

Cooking and other forms of heating of proteinaceous foods leads to the development of flavor compounds due to the Maillard reaction and also to some thermal degradation of cysteine and cystine residues in proteins, as well as of different low-molecular-weight compounds. The essential nitrogenous precursors of the flavor of cooked meat are amino acids, peptides, glycopeptides, and proteins, as well as nucleotides, nucleosides, purines, pyrimidines, and thiamine. Due to thermal degradation of these and some other components, many low-molecular-weight products can be formed. They contribute themselves to the flavor and/or can participate in further reactions, especially with compounds derived from lipids and carbohydrates, as well as with Maillard-reaction products. Therefore, the flavor of cooked muscle is a characteristic of different species of fish. Prell and Sawyer (1988) classified 17 species of North Atlantic fish into 4 groups according to the cooked muscle flavor: with a shellfish, earthy, fish oil–gamey–sour, and fish oil–sour–stronger gamey notes.

The total number of identified volatile flavor components in meat reaction systems is on the order of 700. Among the flavor constituents, identified in different cooked meat, poultry, and fish samples, are mainly sulfur-containing compounds, hydrocarbons, acids, aldehydes, ketones, alcohols, and furane derivatives. They may be present in headspace gases of sterilized foods in quantities ranging from a few to thousands of parts per million, and some of them have an odor-detection threshold as low as 2×10^{-8} g/dm^3 (Schutte, 1974; Shahidi et al., 1986; Bailey and Einig, 1986).

Hydrogen Sulfide

One of the earliest identified volatile flavor compounds produced by heating in muscle foods has been hydrogen sulfide (Frączak and

Pajdowski, 1955; Johnson and Vickery, 1964; Aitken and Connell 1979). It is formed mainly due to thermal decomposition of cysteine/cystine residues in proteins and not from soluble low-molecular-weight compounds of the sarcoplasm because the S–S and SH groups in amino acid residues in proteins make up about 95% of the total of these groups in meats. The generation of H_2S in heated muscle foods increases with the time and temperature of heating and depends on the pH (Johnson and Vickery, 1964). A significant increase in the quantity of generated H_2S takes place at temperatures above 100°C. In the experiments of Opstved et al. (1984), a loss of SH+SS groups in fish meat heated for 20 min was found only above 95°C and amounted to about 10% at 115°C. In squid cooked 45 min at 98°C the decrease in the contents of free SH groups is not accompanied by substantial liberation of H_2S (Synowiecki and Sikorski, 1988). No significant amount of H_2S are formed in cooked seal muscle proteins (Synowiecki and Shahidi, 1991). A large quantity of hydrogen sulfide, however, is generated in oversterilized canned fish and as a result of very long heating. In tuna, about 0.030 mg H_2S/g protein is liberated in 450 min of cooking (Khayat, 1978). Hydrogen sulfide contributes to the flavor of boiled seafoods per se, as well as being a precursor in the formation of other aroma compounds.

Other Volatile Components in Heated Seafoods

One of the components responsible for the characteristic meaty flavor and aroma of canned tuna is 2-methyl–3-furanthiol, which is formed via a thermally mediated reaction between ribose and cysteine (Lindsay, 1990).

About 50% of the almost 100 volatile compounds identified in cooked small shrimp have heterocyclic structure and contain sulfur and/ or nitrogen atoms. The main flavor compounds in cooked shrimp are pyrazines and cyclic polysulfides. In heated shrimp rich in free amino acids, predominantly pyrazines are formed, whereas in species with a lower concentration of free amino acids, mainly trithiolanes and dithiazines are generated (Kubata et al., 1989). The flavor of cooked shrimp is significantly affected by lipoxygenase. The addition of lipoxygenase increases the generation of tetradecatrien–2-one and of the total volatile compounds in heated shrimp (Kuo and Pan, 1991; Pan and Kuo, 1991). Three pyrazines were found in boiled crayfish tail meat. They contribute a nutty, roasted, or toasted character to heated foods. On the other hand, dimethyl sulfide and trimethylsulfide, also found in cooked crayfish meat, are responsible for the undesirable cooked cabbage and spoilage odor (Hsieh et al., 1989).

Table 7–2   Volatile nitrogen- and sulfur-containing compounds in boiled scallop meat

N-Containing Compounds	S-Containing Compounds
Trimethylamine	Methanethiol
1-Methylpyrrole	Dimethylsulfide
Pyrazine	Dimethyl disulfide
Pyridine	Methional
2,6-Dimethylpyrazine	Dimethyl trisulfide
2,3-Dimethylpyrazine	S-Methyl methylthiosulfonate
4,5-Dimethylthiazole	2-Formyl–5-methylthiophene
2,3,5-Trimethylpyrazine	
2-Acetylpyrazine	
2-Acetylpyrrole	
2-Acetyl–3-methylpyrazine	
2,3,5,6-Tetramethylpyrazine	

Source: Adapted from Suzuki et al. (1990).

The role of different precursors and of the parameters of processing is especially important for the flavor of cooked crab. The concentration of several amino acids and nucleotides in the extract of crabmeat fluctuates during heating. Some of them are possibly changed into different heterocyclic compounds responsible for the cooked flavor of crab. By using optimum concentrations of several amino acids and nucleotides, which participate in the heat generation of aroma compounds, and by applying proper conditions of heating, a typical cooked crab flavor should be developed in surimi-based crab analogue products (Hayashi et al., 1990).

Among the 84 volatile compounds identified in the adductor muscle of boiled scallop, a most important flavor component is dimethyl sulfide, followed by breakdown products of polyunsaturated fatty acids and by Maillard-reaction products. The characteristic cooked note of boiled scallop flavor is imparted by N- and S-containing compounds (Table 7–2), whereas TMA generated by thermal decomposition of TMAO contributes to the marine-like flavor (Suzuki et al., 1990).

CHANGES IN NUTRITIONAL VALUE

The effects of heating on the contribution of nitrogenous components of fish to the nutritional value depend on the severity of the treatment applied in different methods of preservation and culinary preparation of seafoods. Generally, because of the characteristic histological structure of fish and marine invertebrates and the low content of collagen, only mild heating is required and the loss of nutritional value is not high (Aitken and Connell, 1979). Opstved *et al.* (1984) found that heating for 20 min at an internal temperature of 95°C, as well as drum-drying at a steam temperature of 145°C, increased the number of disulfide bonds and very slightly decreased protein and amino acid digestibility of fish meat used in feeding rainbow trout. The digestibility of protein and individual amino acids in the cooked samples was on the average only 1.1% and 0.1–2.4% lower, respectively, than in the raw meat. The figures for drum-dried fish were 4.9% and 2.0–11.1%, as compared to those for freeze-dried samples. The effect of heating was higher in pollock than in mackerel flesh.

In seafood products heated under conditions favoring the formation of different Maillard products, a loss of nutritive value may result from the involvement of essential amino acids in these reactions (Dworschak, 1980). Frying fish of different species for 4 min in fresh oil at 180°C may decrease the FDNB-available lysine by about 18%. When highly oxidized oil is used, the loss is about 29% (Tooley and Lawrie, 1974). In model experiments, the loss of available lysine in water-soluble and salt-soluble proteins of mackerel due to heating at 80, 100, and 120°C was very small, the highest being 15% in samples heated as long as 240 min at 120°C. It increased with the addition of reducing sugars and at higher temperature and pH (Tanaka et al., 1980, 1981).

The effect of heat on nitrogenous components of seafoods may have also a food safety aspect because heating may serve to inactivate marine toxins. The kinetics of heat inactivation of paralytic shellfish poison in soft-shell clams was found to be similar to those of most microorganisms (Gill et al., 1985).

REFERENCES

AITKEN, A., and CONNELL, J. J. 1979. "Fish." Pp. 219–253. In *Effects of Heating on Foodstuffs,* edited by R. J. Priestley. London: Applied Science Publishers.

BABBITT, J. K. 1982. "Blueing Discoloration of Dungeness Crabmeat." Pp. 423–428. In *Chemistry and Biochemistry of Marine Food Products,* edited by R. E. Martin, G. J. Flick, C. E. Hebard, and D. R. Ward. New York: Van Nostrand Reinhold.

BAILEY, M. E., and EINIG, R. G. 1986. "Reaction Flavors of Meat." In *Thermal Generation of Aromas,* edited by T. H. Parliment, R. J. McGorrin, and Ch. T. Ho. Chap. 39. Washington, DC: American Chemical Society.

BEAS, V. E., WAGNER, J. R., CRUPKIN, M., and ANON, M. C. 1990. "Thermal Denaturation of Hake (*Merluccius hubbsi*) Myofibrillar Proteins. A Differential Scanning Calorimetric and Electrophoretic Study." *J. Food Sci.* 55:683–687.

BERAQUET, N. J., MANN, J., and AITKEN, A. 1984. "Heat Processing of Herring. I. Release of Water and Oil." *J. Food Technol.* 19:437–446.

BOISMENU, D., LEPINE, F., GAGNON, M., and DUGAS, H. 1990. "Heat Inactivation of Catalase from Cod Muscle and from Some Psychrophilic Bacteria." *J. Food Sci.* 55:581–582.

BOON, D. D. 1977. "Coloration in Bivalves. A Review." *J. Food Sci.* 42:1008–1015.

BUTTKUS, H. 1974. "On the Nature of the Chemical and Physical Bonds Which Contribute to Some Structural Properties of Protein Foods. A Hypothesis." *J. Food Sci.* 39:484–489.

BYKOWSKI, P., and KOŁODZIEJSKI, W. 1983. "Właściwości Mięsa z Kryla Odskorpionego Metodą Rolkową." *Bull. Sea Fish. Instit. Gdynia* 14 (5–6):53–57.

DOWDIE, O. G., and BIEDE, S. L. 1983. "Influence of Processing Temperature on the Distribution of Tissue and Water-Soluble Proteins in Blue Crabs (*Callinectes sapidus*)." *J. Food Sci.* 48:804–807, 812.

DUNAJSKI, E. 1977. "Wpływ Temperatury, pH i Siły Jonowej na Twardość Mięśni Ryb." P. 22. 8 Sesja Naukowa KTCHZ PAN, Streszczenia Doniesień. Poznań. Komitet Technologii i Chemii Żywności PAN.

DUNAJSKI, E. 1979. "Texture of Fish Muscle." *J. Texture Studies* 10:301–318.

DWORSCHAK, E. 1980. "Nonenzyme Browning and Its Effect on Protein Nutrition." *Crit. Rev. Food Sci. Nutr.* 13:1–40.

PRĄCZAK, A., and PAJDOWSKI, Z. 1955. "O Rozkładzie Grup Sulfhydrylowych pod Wpływem Obróbki Termicznej w Mięsie." *Przemysł Spożywczy.* 9:334–336.

GILL, T. A., THOMPSON, J. W., and GOULD, S. 1985. "Thermal Resistance of Paralytic Shellfish Poison in Soft-Shell Clams." *J. Food Protect.* 48:659–662.

HAMANN, D. D., AMATO, D. M., WU, M. C., and FOEGEDING, E. A. 1990. "Inhibition of Modori (Gel Weakening) in Surimi by Plasma Hydrolysate and Egg White." *J. Food Sci.* 55:665–669.

HAYASHI, T., ISHII, H., and SHINOHARA, A. 1990. "Novel Model Experiment for Cooking Flavor Research on Crab Leg Meat." *Food Rev. Internat.* 6:521–536.

HSIEH, T. C. Y., VEJAPHAN, W., WILLIAMS, S. S., and MATIELLA, J. E. 1989. "Volatile Flavor Compounds in Thermally Processed Louisiana Red Swamp Crayfish and Blue Crab." In *Thermal Generation of Aromas*, edited by T. H. Parliment, R. J. McGorrin, and C. T. Ho. ACS Symposium Series 409. Chap. 36. Washington, DC: American Chemical Society.

JOHNSON, E. R., and VICKERY, J. R. 1964. "Factors Influencing the Production of Hydrogen Sulphide from Meat During Heating." *J. Sci. Food Agric.* 15:695–701.

KANOH, S., POLO, J. M. A., KARIYA, Y., KANEKO, T., WATABE, S., and HASHIMOTO,

K. 1988. "Heat-induced Textural and Histological Changes of Ordinary and Dark Muscles of Yellowfin Tuna." *J. Food Sci.* 53:673–678.

KHAYAT, A. 1978. "Hydrogen Sulfide Production by Heating Different Protein Fractions of Tuna Meat." *J. Food Biochem.* 2:121–131.

KINOSHITA, M., TOYOHARA, H., and SHIMIZU, Y. 1990. "Characterization of Two Distinct Latent Proteinases Associated with Myofibrils of Crucian Carp (*Carassius auratus cuvieri*)." *Comp. Biochem. Physiol.* 97B:315–319.

KREUZER, R. 1984. "Cephalopods: Handling, Processing, and Products." FAO Fish Tech. Paper 254. Rome: Food and Agriculture Organization.

KUBATA, K., UCHIDA, CH., KUROSAWA, K., KOMURO, A., and KOBAYASHI, A. 1989. "Identification and Formation of Characteristic Volatile Compounds from Cooked Shrimp." In *Thermal Generation of Aromas*, edited by T. H. Parliment, R. J. McGorrin, and C. T. Ho. ACS Symposium Series 409. Chap. 35. Washington, DC: American Chemical Society.

KUO, J. M., and PAN, B. S. 1991. "Effects of Lipoxygenase on Formation of the Cooked Shrimp Flavor Compound 5,8,11–tetradecatrien–2-one." *Agric. Biol. Chem.* 55:847–848.

LALL, B. S., MANZER, A. R., and HILTZ, D. F. 1975. "Treatment for Improvement of Frozen Storage Stability at $-10°C$ in Fillets and Minced Flesh of Silver Hake." *J. Fish. Res. Bd. Canada* 32:1450–1454.

LINDSAY, R. C. 1990. "Fish Flavors." *Food Rev. Internat.* 6:437–455.

MOTOHIRO, T. 1982. "The Effect of Heat Processing on Color Characteristics in Crustacean Blood." Pp. 405–413. In *Chemistry and Biochemistry of Marine Food Products*, edited by R. E. Martin, G. J. Flick, C. E. Hebard, and D. R. Ward, New York: Van Nostrand Reinhold.

NAKAGAWA, T., and NAGAYAMA, F. 1989. "Interaction of Fish Muscle Glycolytic Enzymes with F-actin and Actomyosin." *Nippon Suisan Gakkaishi.* 55:165–171.

OPSTVED, J., MILLER, R., HARDY, R. W., and SPINELLI, J. 1984. "Heat-induced Changes in Sulfhydryl Groups and Disulfide Bonds in Fish Protein and Their Effect on Protein and Amino Acid Digestibility in Rainbow Trout (*Salmo gairdneri*)." *J. Agric. Food Chem.* 32:929–935.

OTWELL, W. S., and HAMANN, D. D. 1979. "Textural Characterization of Squid *Loligo pealei*." *J. Food Sci.* 44:1629–1635, 1643.

PAN, B. S. 1990. "Minced Fish Technology." Pp. 190–210. In *Seafood: Resources, Nutritional Composition, and Preservation*, edited by Zdzisław E. Sikorski. Boca Raton, FL: CRC Press.

PAN, B. S., and KUO, J. M. 1991. "Flavor of Shellfish and Kamaboko Flavorants." Paper read at 8th World Congress of Food Science and Technology, 29 September–4 October, Toronto, Canada.

PAN, B. S., KUO, J. M., LUO, L. J., and YANG, H. M. 1986. "Effect of Endogenous Proteinases on Histamine and Honeycomb Formation in Mackerel." *J. Food Biochem.* 10:305–319.

PRELL, P. A., and SAWYER, F. M. 1988. "Flavor Profiles of 17 Species of North Atlantic Fish." *J. Food Sci.* 53:1036–1042.

SADOWSKA, M., and SIKORSKI, Z. E. 1977. "Evaluation of Technological Suitability of Fish Meat by Rheological Measurements." *Lebens.-Wiss. u.-Technol.* 10:239–245.

SCHUTTE, L. 1974. "Precursors of Sulfur-Containing Flavor Compounds." *Crit. Rev. Food Technol.* 4:457–504.

SHAHIDI, F., RUBIN, L. J. and D'SOUZA, L. A. 1986. "Meat Flavor Volatiles: A Review of the Composition, Techniques of Analysis, and Sensory Evaluation." *Crit. Rev. Food Sci. Nutr.* 24:141–243.

SIKORSKI, Z. E., and KOŁODZIEJSKA, I. 1986. "The Composition and Properties of Squid Meat." *Food Chem.* 20:213–224.

SIKORSKI, Z. E., SCOTT, D. N., and BUISSON, D. H. 1984. "The Role of Collagen in the Quality and Processing of Fish." *Crit. Rev. Food Sci. Nutr.* 20:301–343.

SPINELLI, J., and KOURY, B. 1979. "Nonenzymic Formation of Dimethylamine in Dried Fishery Products." *J. Agric. Food Chem.* 27:1104–1108.

SPINELLI, J., and KOURY, B. 1981. "Some New Observations on the Pathways of Formation of Dimethylamine in Fish Muscle and Liver." *J. Agric. Food Chem.* 29:327–331.

STANLEY, D. W., and HULTIN, H. O. 1982. "Quality Factors in Cooked North Atlantic Squid." *Can. Inst. Food Sci. Technol. J.* 15:227–282.

SUZUKI, J., ICHIMURA, N., and ETOH, T. 1990. "Volatile Components in Boiled Scalop." *Food Rev. Internat.* 6:537–552.

SYNOWIECKI, J., and SIKORSKI, Z. E. 1988. "Heat Induced Changes in Thiol Groups in Squid Proteins." *J. Food Biochem.* 12:127–135.

SYNOWIECKI, J., and SHAHIDI, F. 1991. "Heat-induced Changes in Sulfhydryl Groups of Seal Muscle Protein." *J. Agric. Food Chem.* 39:2006–2009.

TANAKA, M., OKUBO, S., SUZUKI, K., and TAGUCHI, T. 1980. "Available Lysine Losses in Water Soluble Protein of Mackerel Meat by Heating." *Bull. Japan. Soc. Sci. Fish.* 46:1539–1543.

TANAKA, M., OKUBO, S., SUZUKI, K., and TAGUCHI, T. 1981. "Available Lysine Losses in Salt Soluble Proteins of Mackerel by Heating." *Bull. Japan. Soc. Sci. Fish.* 47:1075–1078.

TOOLEY, P. J., LAWRIE, R. A. 1974. "Effect of Deep Fat Frying on the Availability of Lysine in Fish Fillets." *J. Food Technol.* 9:247–253.

TOYOHARA, H., KINOSHITA, M., and SHIMIZU, Y. 1990. "Proteolytic Degradation in Threadfin-Bream Meat Gel." *J. Food Sci.* 55:259–260.

TOYOHARA, H., SAKATA, T., YAMASHITA, K., KIMOSHITA, M., and SHIMIZU, Y. 1990a. "Degradation of Oval-Filefish Meat Gel Caused by Myofibrillar Proteinase(s)." *J. Food Sci.* 55:364–368.

TOYOHARA, H., SUSAKI, K., KINOSHITA, M., and SHIMIZU, Y. 1990b. "Effect of Bleeding in the Modori Phenomenon and Possible Existence of Some Modori-Inhibitor in Serum." *Nippon Suisan Gakkaishi.* 56:1245–1249.

ZAITSEV, V., KIZEVETTER, I., LAGUNOV, L., MAKAROVA, T., MINDER, L., and PODSEVALOV, V. 1969. *Fish Curing and Processing.* Moscow: Mir Publishers.

8

CHANGES IN PROTEINS IN FROZEN STORED FISH

Zdzisław E. Sikorski and Anna Kołakowska

INTRODUCTION

Frozen fish stored several months at about $-20°C$ may, after cooking, become tough, chewy, rubbery, stringy, or fibrous. This is accompanied by a loss in functional characteristics of the muscle proteins, mainly solubility, water retention, gelling ability, and lipid emulsifying properties. Freezing and thawing may result in lysis of mitochondria and lysosomes and a change in distribution of enzymes (Karvinen et al., 1982). A gradual decline in the activities of various muscle enzymes has also been observed during storage at freezing temperatures. The loss of ATPase activity, both in meat homogenates and in protein solutions, may reach 50–80% (Buttkus, 1967). The total solubility of proteins in neutral 5% NaCl solution may decrease to about 30%, whereby the main loss regards the contractile proteins, mainly myosin heavy chain, M-proteins, tropomyosin, and troponins I and C in descending order (Owusu-Ansah and Hultin, 1992). It was shown by Jarenbäck and Liljemark (1975b), that the myosin microfibrils in fresh cod muscle could be almost totally extracted, whereas myosin in frozen muscle after prolonged storage was resistant to extraction. Significant changes in the sodium dodecyl sulfate–polyacrylamide gel electrophoresis (SDS—PAGE) and HPLC profiles of cod sarcoplasmic proteins due to frozen storage have been shown recently by LeBlanc and LeBlanc (1992b).

These changes are caused by processes known as freeze denaturation of proteins, involving usually denaturation proper, followed by interactions of the denatured proteins with various reactive components of the fish tissues. Fish muscle proteins, mainly myosin, are more susceptible to abuse conditions of freezing and frozen storage than those from land animals.

Among the factors involved in freeze denaturation are the effects of ice crystals and ions, binding of fatty acids and of lipid oxydation products to proteins, oxidation and interactions of thiol groups, as well as the chemical reactions of amino acid residues in proteins with endogenous formaldehyde (FA) and other reactive components in the muscles (Love, 1966; Sikorski et al., 1976; Matsumoto, 1979; Shenouda, 1980; Suzuki, 1981; Sikorski and Kołakowska, 1990; Haard, 1992).

Freeze denaturation of proteins in fish of different species is affected by the content and distribution of fat in the tissues, as well as by the rate of accumulation of FA due to demethylation of TMAO, of different amino acids, and of the products of nucleotide catabolism. The technological factors affecting freeze denaturation comprise the kind of processing before freezing, that is, filleting, skinning, mincing, and washing of the mince. Generally, the protein changes are more pronounced in extracted proteins stored in the form of solutions or suspensions than in the intact muscle. This is especially evident in the case of squid proteins (Iguchi et al., 1981). Washing of the mince with water prior to freezing may lessen toughening, possibly by partially inhibiting the enzymatic formation of FA. Changing the water content in the fish mince may affect the role of ice in protein denaturation (Boer and Fennema, 1989). Further important factors are the rate of freezing, the time and temperature of storage, and the effectiveness of glazing or protective packaging (Santos and Regenstein, 1990). Recently, the effect of pressure processing (Leblanc et al., 1987) as well as of glazing and of protective and vacuum packaging has been investigated (Santos and Regenstein, 1990; Sirois et al., 1991).

THE EFFECT OF FREEZING AND STORAGE CONDITIONS

The rate of freezing and the temperature of storage affect the size and distribution of ice crystals in the tissues and may change the microstructure of fish muscle. This has been shown already in the early investigations of Plank et al. (1916), as well as in many later papers, for example, by Jahrenbäck and Liljemark (1975a) and Lampila and Brown (1986). The rate of freezing also determines how long the frozen fish is exposed to the abusing temperature range of about −1 to −5°C, at which point many catabolic reactions in the muscle proceed faster than in the unfrozen tissue. Freezing brings also about a change in the hydration of ions. As a consequence, there is a redistribution of the hydrogen bonds and hydrophobic adherences in the proteins. Crystallization of ice may disrupt the water structures contributing to the hydrophobic adherences which participate in buttressing the native protein conformation (Lewin, 1974). Furthermore, the loss of fluid water may favor protein aggregations by increasing the concentration of reactive solutes. The rate of aggregation of trout myosin is higher at −10°C, that is, when most of the tissue water is frozen, than at 0, −20, and −30°C (Buttkus, 1970).

Different ions have the characteristic ability to interfere with the water structures and, thus, to affect the protein conformation in solutions. The multivalent cations La^{3+}, Mg^{2+}, Ba^{2+}, and Ca^{2+} have the highest capacity for breaking the water structures. At a rather low concentration, corresponding to ionic strength of 0.5–1, many inorganic salts increase the protein solubility by rupturing the ionic bonds between the amino acid residues and by forming hydrated protein–ion associations. At higher concentrations, the competition of the ions and of the hydrophilic protein groups for water may lead to a decrease in protein hydration, resulting in loss of solubility. Furthermore, many salts increase the surface tension of solutions, favoring the formation of further intramolecular hydrophobic adherences. Ca^{2+} and Mg^{2+} at the appropriate pH may affect the protein conformation directly by forming cross-links between ionized amino acid residues. Many inorganic salts may also contribute to the changes in proteins by catalyzing lipid hydrolysis and autoxidation, participating in the formation of lipid–protein complexes, and promoting the oxidation of thiol groups (Kołodziejska and Sikorski, 1980).

The results of numerous model experiments with isolated myofibrillar proteins as well as extensive industrial experience indicate that the stability of fish muscle proteins increases very significantly with a decrease in storage temperature from −20 to −40°C (Jiang et al., 1989). To ensure the high quality of valuable fish frozen at sea, the catch may be stored on fishing vessels even at temperature as low as −50°C.

THE EFFECT OF LIPIDS

The role of lipids in freezing changes in proteins has been investigated and reviewed during the last 30 years (Dyer and Dingle, 1961; Sikorski et al., 1976; Shenouda, 1980; Matsumoto, 1979; Sikorski and Kołakowska, 1990; Haard, 1992). Lipids, especially oxidized lipids, may affect the hydrogen bonds and hydrophobic adherences in the proteins of frozen fish. The carbonyl groups of oxidized lipids may participate in covalent bonding, leading to the formation of stable protein–lipid aggregates. The role of lipids in freezing denaturation of proteins depends on the quantity and distribution of the lipids in the tissues, their properties, and their hydrolysis and oxidation.

The contents of lipids in the tissues is affected by the fish species, size, age, and sex, by the fishing period and many other biological factors, and also by the type of tissue and location in the animal body (Sikorski et al., 1990). The lipid-to-protein ratio may range from 0.1 to above 2. Paradoxically enough, the effect of lipids on protein changes in frozen fish is most significant in lean species. In the lean fish muscle, the lipids are limited to the physiologically necessary membrane lipids, that is, they are comprised of phospholipids almost solely and a little amount of sterol esters. Hydrolysis and oxidation of these lipids may result in membrane damage and increased membrane permeability. This, in consequence, may lead to increased activity of enzymes directly or indirectly involved in protein changes.

The hypotheses of participation of lipids in freeze denaturation of proteins are based on:

- Correlation between protein and lipid changes in fish tissues during freezing and frozen storage
- Evidence of several factors affecting both lipid hydrolysis/oxidation and protein denaturation in frozen fish
- Results of model experiments on protein–lipid interactions

The correlation between lipid and protein changes in frozen fish gives circumstantial evidence to the hypotheses. Since the time of Dyer's review, further reports confirmed the correlation between the increase in the contents of free fatty acids in the muscles of frozen fish and the decrease in protein extractability in salt solutions. Such a relationship was observed for many lean and fatty species stored under various conditions. Both the soluble protein and the free fatty acid content have been used as quality criteria of frozen fish (Konig and Mol, 1991). A close correlation was found between the extractability of salt-soluble proteins and malonaldehyde content, peroxide value, iodine value, and other indices of lipid oxidation in frozen fish, mainly in fatty species. The results of recent studies, however, do not indicate a significant role of lipid oxidation products in freezing denaturation of proteins in lean fish. Although Knudsen et al. (1990) demonstrated a relationship between the gel-forming ability and lipid hydrolysis in cod fillets and mince stored at −20°C, their results do not show a clear effect of lipid oxidation or the addition of pro- and antioxidants. Rehbein and Orlick (1990) found that even severe lipid oxidation, up to 14 mg of malonaldehyde per kg w/w, had only a slight effect of the texture of frozen minced fillets of Antarctic fish. In frozen stored fillets of Baltic herring from various fishing seasons, a correlation was found between the changes in elastic moduli and the peroxide value. The salt-soluble protein extractability, however, was better correlated with the TBA value and the contents of free fatty acids than with the peroxide value or texture in fatty and lean herring (Kołakowska et al., 1992).

It should not be expected, however, to find in all published papers high correlations between protein and lipid changes in frozen fish, although the potential of interactions of lipids and lipid oxidation products with proteins is known. Not all the methods used for measuring protein denaturation, and a variety of tests of lipid oxidation applied in different experiments, reflect strictly the changes induced in proteins by interactions with lipids. The commonly used TBA test gives positive results even in the absence of lipids (Kołakowska and Deutry, 1983). Furthermore, a complete recovery of lipid oxidation products and of fatty acids after lipid–protein interactions is not possible.

Protein–lipid bonds are present naturally in fish tissues. Thus, it is necessary to use nonpolar and polar solvents for exhaustive extraction of lipids from fish flesh. For recovering some lipids, even hydrolysis of the material prior to extraction is necessary. These covalently bonded fractions contains all classes of lipids (Kołakowska and Deutry, 1984). Changes in the amounts of bound lipids during frozen storage depend on the species of fish and conditions of storage. In Baltic herring stored 9 months at −28°C, no significant changes in the amount of covalently bonded lipids were detected (Czerniejewska-Surma, 1985). However, in horse mackerel stored 3–14 months at the same temperature, the proportion of bonded to total lipids increased linearly with the storage time (Kołakowska et al., 1985).

In model systems, the interaction of proteins with polar and neutral lipids, free fatty acids, and the products of all stages of lipid oxidation has been demonstrated. The higher activity of lipid hydroperoxides and, generally, of oxidized lipids than of free fatty acids was shown by Jarenbäck and Liljemark (1975c) and Kołakowska (1979), respectively. These interactions, however, are affected by the presence of different classes of lipids (Takama et al., 1972).

THE INTERACTION WITH FORMALDEHYDE

The differences in the rate of freeze denaturation of proteins in fish belonging to various species may be caused by the action of FA, generated together with DMA as the result of enzymatic decomposition of TMAO (Sikorski and Kostuch, 1982; Lundstrom et al., 1982). A large accumulation of DMA and FA in the muscles of frozen fish is generally accompanied by a decrease in extractability of myofibrillar proteins (Tokunaga, 1964). Being a very reactive compound, FA is capable of reacting with many protein functional groups, thereby changing their effect on the protein conformation, or may induce the formation of cross-links. The addition of FA to fish meat brings about a significant reduction of protein solubility (Sikorski et al., 1975). By binding to the functional groups of amino acid residues in fish muscle proteins, FA may cause the symptoms of denaturation, measured as a loss in extractability and enzymatic activity, as well as change in hydrophobicity, also without causing new cross-links. According to Ang and Hultin (1989) and Hultin (1992), FA could not only induce toughening in frozen stored fish by direct cross-linking of the proteins but also, causing denaturation of the proteins by binding to their side chain groups, it could lead to increased formation of aggregates, buttressed by noncovalent forces. On the other hand, the loss of TMAO due to degradation to DMA and FA may itself contribute to increased susceptibility of fish proteins to freezing denaturation (Owsu-Ansah and Hultin, 1984).

Regardless of the rate of accumulation of DMA and the corresponding generation of FA in the fish meat, there is a high correlation between the concentration of DMA and protein extractability. At DMA contents of about 6.4–7.0 mg/100 g and a corresponding amount of generated FA, the yield of extractable myofibrillar proteins decreases by about 50% of that in fresh cod (Kostuch, 1978). It was shown in model experiments that increasing the pH of cod mince favors the binding of FA to proteins both at 4 and −18°C. However, at a higher pH the loss in protein extractability is lower than in a more acidic environment, probably due to less cross-linking. An increase in ionic strength in the mince favors the reactions with FA leading to the decrease in protein solubility at 4°C (Kostuch and Sikorski, 1977). In gadoid fish meat stored at −7°C, the generation of FA has been found to be accompanied by formation of a cross-linked protein of 280,000 Da (Ragnarsson and Regenstein, 1989). Changes in the molecular weight of proteins of the myofibrillar and sarcoplasmic fractions in frozen stored cod were shown recently by LeBlanc and LeBlanc (1992). Heating of frozen stored squid mantle brings about cross-linking of myofibrillar proteins, as evidenced in the electrophoretic pattern. Such changes could not be found in the proteins of fresh squid (Nitisewojo, 1987).

The generation of FA and the decrease in protein extractability are more rapid in fish minces than in gutted fish or fillets. By extracting the water-soluble fractions from fish meat before freezing, the extent of DMA accumulation in the frozen product can be reduced.

THE ROLE OF DISULFIDE BONDS

Oxidation of thiol groups has long been regarded as possibly being involved in the denaturation of proteins in frozen fish. In the early investigations, no significant changes in the number of SH groups in the muscles of frozen stored fish could be evidenced (Connell, 1968; Mao and Sterling, 1970). However, cross-linking of proteins due to disulfide bond formation could take place without a significant net change in the total −SH groups by SH–disulfide exchange reactions (Buttkus, 1970). Furthermore, recently, Chen et al. (1989) reported on significant decrease in the total SH content of milkfish myosin at −20°C, and Jiang et al. (1988a, 1988b) showed that denaturation of myosin in milkfish actomyosin during freezing and storage at −20°C was accompanied by the formation of disulfide bonds. In actomyosin extracted from tilapia hybrid, the total contents of SH groups decreased significantly during freezing and subsequent storage at −20°C, whereas at −40°C, the change was slight during the first 2 weeks and insignificant later (Jiang et al., 1989). Jiang et al. (1991) found that significant loss of total SH groups in actomyosin during frozen storage of grass prawn was correlated to the decrease in Ca–ATPase activity.

THE INTERACTION OF DIFFERENT FACTORS

The results of many investigations reveal that different factors contribute to the changes in proteins in frozen stored fish. A very interesting discussion of the role of hydrophobic interactions, FA, and disulfide bonds in these changes has been presented by Owusu-Ansah and Hultin (1987). It appears that the decrease in extractability of salt-soluble proteins is primarily caused by formation of new hydrogen bonds and hydrophobic adherences. After several months at −13°C, most of the myofibrillar proteins in fish meat can still be extracted by SDS. However, the possibility of the formation of different covalent cross-links in the proteins of frozen fish cannot be excluded. In fish known to accumulate large amounts of

DMA, the SDS-soluble protein content is higher in the inner than in the outer dehydrated and oxidized layers of muscle (Sikorski, 1980). The addition of reducing substances may significantly increase the extractability of the proteins of frozen stored fish (Owusu-Ansah and Hultin, 1986). It was shown recently that $NaBH_4$ increased the solubility of actomyosin denatured during frozen storage (Jiang et al., 1988b). The formation of a few bonds resistant to SDS must not necessarily lead to a significant decrease in protein solubility. On the other hand, even slight cross-linking may induce conformational changes, favoring the formation of additional hydrophobic adherences and hydrogen bonds, which may lead to protein aggregation and loss in solubility. Deconformation, resulting from changes in the original noncovalent bonds in the proteins, may promote the interactions of amino acid residues with reactive muscle components. The involvement of some functional groups in proteins in these interactions does not necessarily need to be reflected in their net loss, as was shown by Buttkus (1970) regarding the SH groups of myosin. LeBlanc and LeBlanc (1992a) observed recently in different protein fractions of frozen cod, stored 1 year at temperatures from -12 to $-30°C$, significant changes in SH, S–S, and free NH_2 groups, aldehydes, ester-links, and surface hydrophobicity and discussed the role of these changes in freezing denaturation of fish proteins.

The proteins in frozen stored fish may be affected by reactions with different metabolites formed in the muscles postmortem. In model experiments, ADP, AMP, and IMP inhibit protein denaturation during frozen storage, whereas Ino and Hx are not active or else increase the rate of changes (Jiang et al., 1987a, 1987b). Fish muscles containing large amounts of free glycine, proline, and alanine are more resistant to freeze denaturation than those rich in free histidine and taurine (Jiang and Lee, 1985).

PROTECTION AGAINST FREEZE DENATURATION

The deteriorative protein changes in frozen fish can be minimized or prevented in whole fish or fillets by applying a very low, constant temperature of storage and by inhibiting oxidation of lipids and desiccation of the product. In fish minces, additional effects can be achieved by removing the water-soluble fraction containing both TMAO and the factors catalyzing its degradation to FA and by adding cryoprotectants.

The effectiveness and mode of action of different protectants in model systems and in frozen fish minces was studied by Matsumoto and co-workers (Noguchi and Matsumoto, 1970; Matsumoto and Noguchi, 1992) as well as by other groups (Tran, 1974; Kołodziejska et al., 1989; Krueger and Fennema, 1989; Yoon and Lee, 1990). Several sugars, polyphosphates, amino acids, peptides, carboxylic acids, α-ketoacids, and polyalcohols were found very effective. The principles of action of different types of cryoprotectants has been discussed by Lanier in this volume. An effective novel approach was proposed by Jiang et al. (1986), based on the effects of reducing agents on protein denaturation in frozen minced fish (Lan, 1987). By adding different reductants during the early stages of grinding of freeze-thawed cod and mackerel meat, the –SH groups were liberated. After adjustment of the pH to neutral, the addition of oxidants during the final stage of grinding induced reformation of S–S bonds and improved the structure of kamaboko.

REFERENCES

ANG, J. F., and HULTIN, H. O. 1989. "Denaturation of Cod Myosin During Freezing After Modification with Formaldehyde." *J. Food Sci.* 54:814–818.

BOER, G., and FENNEMA, O. 1989. "Effect of Mixing and Moisture Modification on Toughening and Dimethylamine Formation in Alaska Pollock Mince During Frozen Storage at −10°C." *J. Food Sci.* 54:1524–1529.

BUTTKUS, H. 1967. "The Reaction of Myosin with Malonaldehyde." *J. Food Sci.* 32:432–434.

BUTTKUS, H. 1970. "Accelerated Denaturation of Myosin in Frozen Solution." *J. Food Sci.* 35:558–562.

CHEN, C. S., HWANG, D. C., and JIANG, S. T. 1989. "Effect of Storage Temperatures on the Formation of Disulfides and Denaturation of Milkfish Myosin (*Chanos chanos*)." *J. Agric. Food Chem.* 37:1228–1231.

CONNELL, J. J. 1968. "Effect of Freezing and Frozen Storage on the Proteins of Fish Muscle." Pp. 333–358. In *Low Temperature Biology of Foodstuffs*. Edited by J. Hawthorn and E. J. Rolfe. Oxford: Pergamon Press.

CZERNIEJEWSKA-SURMA, B. 1985. *Zależność Między Zjełczeniem Mrożonych Śledzi Bałtyckich i Wyprodukowanych z nich Konserw.* Thesis. Akademia Rolnicza, Szczecin.

DYER, W. J. and DINGLE, J. R. 1961. "Fish proteins with Special Reference to Freezing." Pp. 275–327. In *Fish as Food,* edited by George Borgstrom. Vol. 1. New York: Academic Press.

HAARD, N. F. 1992. "Biochemical Reactions in Fish Muscle During Frozen Storage." Pp. 176–209. In *Seafood Science and Technology,* edited by G. Bligh. Oxford: Fishing News Books.

HULTIN, H. O. 1992. "Trimethylamine–N–Oxide (TMAO) demethylation and protein denaturation in fish muscle." Pp. 25–42. In *Advances in Seafood*

Biochemistry, Composition and Quality, edited by G. J. Flick and R. E. Martin. Lancaster–Basel: Technomic Publishing Co.

IGUCHI, S. M. M., TSUCHIYA, T., and MATSUMOTO, J. J. 1981. "Studies on the Freeze Denaturation of Squid Actomyosin." *Bull. Japan. Soc. Sci. Fish.* 47:1499–1506.

JARENBACK, L., and LILJEMARK, A. 1975a. "Ultrastructural Changes During Frozen Storage of Cod (*Gadus morhua* L.). I. Structure of Myofibrils as Revealed by Freeze Etching Preparation." *J. Food Technol.* 10:229–239.

JARENBACK, L., and LILJEMARK, A. 1975b. "Ultrastructural Changes During Frozen Storage of Cod (*Gadus morhua* L.). II. Structure of Extracted Myofibrillar Proteins and Myofibril Residues." *J. Food Technol.* 10:309–325.

JAHRENBACK, L., and LILJEMARK, A. 1975c. "Ultrastructural Changes During Frozen Storage of Cod (*Gadus morhua* L.). III. Effect of Linoleic Acid and Linoleic Acid Hydroperoxides on Myofibrillar Proteins." *J. Food Technol.* 10:437–552.

JIANG, S. T., and LEE, T. C. 1985. "Changes in Free Amino Acids and Protein Denaturation of Fish Muscle During Frozen Storage." *J. Agric. Food Chem.* 33:839–843.

JIANG, S. T., HWANG, D. C., and CHEN, C. S. 1988a. "Denaturation and Change in SH Group of Actomyosin from Milkfish (*Chanos*) During Storage at $-20°C$." *J. Agric. Food Chem.* 36:433–437.

JIANG, S. T., HWANG, D. C., and CHEN, C. S. 1988b. "Effect of Storage Temperatures on the Formation of Disulfides and Denaturation of Milkfish Actomyosin (*Chanos*)." *J. Food Sci.* 53:1333–1335.

JIANG, S. T., HWANG, B. S., and TSAO, C. Y. 1987a. "Effect of Adenosine Nucleotides and Their Derivatives on the Denaturation of Myofibrillar Proteins in Vitro During Frozen Storage at $-20°C$. *J. Food Sci.* 52:117–123.

JIANG, S. T., HWANG, B. S., and TSAO, C. Y. 1987b. "Protein Denaturation and Changes in Nucleotides of Fish Muscles During Frozen Storage." *J. Agric. Food Chem.* 35:22–27.

JIANG, S. T., LAN, C. C., and TSAO, C. Y. 1986. "New Approach to Improve the Quality of Minced Fish Products from Freeze-Thawed Cod and Mackerel." *J. Food Sci.* 51:310–312.

JIANG, S. T., SAN, P. C., and JAPIT, L. S. 1989. "Effect of Storage Temperatures on the Formation of Disulfides and Denaturation of Tilapia Hybrid Actomyosin (*Tilapia nilotica* × *Tilapia aurea*)." *J. Agric. Food Chem.* 37:633–636.

JIANG, S. T., HWANG, B. S., MOODY, M. W., and CHEN, H. C. 1991. "Thermostability and Freeze Denaturation of Grass Prawn (*Penaeus monodon*) Muscle Proteins." *J. Agric. Food Chem.* 39:1998–2001.

KARVINEN, V. P., BAMFORD, D. H., and GRANROTH, B. 1982. "Changes in Muscle Subcellular Fractions of Baltic Herring (*Clupea harengus membras*) During Cold and Frozen Storage." *J. Sci. Food Agric.* 33:763–770.

KNUDSEN, L., NIELSEN, J., and BORESSEN, T. 1990. "The Effect of Lipid Oxidation and Hydrolysis on Functional Properties of Frozen Cod Mince." Pp. 311–316. In *Chilling and Freezing of New Fish Products*. Proceedings I.I.F.–I.I.R. Commission C2, Aberdeen: International Institute of Refrigeration.

KOŁAKOWSKA, A. 1979. *"Wpływ Wybranych Czynników Technologicznych na Jełczenie Ryb Mrożonych."* Rozprawy No. 67. Szczecin: Akademia Rolnicza.

KOŁAKOWSKA, A., and DEUTRY, J. 1983. "Some Comments on the Usefulness of Rancidity Tests in Frozen Fish." *Nahrung.* 27:513–518.

KOŁAKOWSKA, A., and DEUTRY, J. 1984. "Uwagi o Przydatności Metody Weibulla-Stoldta do Ekstrakcji Związanej Frakcji Lipidów z Mięsa Ryb." *Zeszyty Naukowe Akademii Rolniczej w Szczecinie.* 108:85–94.

KOŁAKOWSKA, A., CZERNIEJEWSKA-SURMA, B., and DEUTRY, J. 1985. "Evaluation of Rancidity in Frozen Horse Mackerel." Pp. 163–167. Proceedings of the 2nd Symposium on Fat Chemistry and Technology, 15–17 September, Gdańsk. Proceedings, Pp. 163–7.

KOŁAKOWSKA, A., KWIATKOWSKA, L., LACHOWICZ, K., GAJOWIECKI, L., and BORTNOWSKA, G. 1992. "Effect of Fishing Season on Frozen Storage Quality of Baltic Herring. Pp. 269–277. In *Seafood Science and Technology*, edited by G. Bligh. Oxford: Fishing News Books.

KOŁODZIEJSKA, I., and SIKORSKI, Z. E. 1980. "Inorganic Salts and Extractability of Fresh and Frozen Fish Proteins." *Internat. J. Refrig.* 3:151–155.

KOŁODZIEJSKA, I., SADOWSKA, M., KOŁODZIEJ, K., and SIKORSKI, Z. E. 1989. "The Effect of Stabilizers on the Functional Properties of Cod, Blue Whiting, and Hake Surimi." Pp. 315–324. In *Engineering and Food. Vol. 1. Physical Properties and Process Control*, edited by W. E. L. Spiess and H. Schubert. London/New York: Elsevier Applied Science.

KONIG, A. J., and MOL, T. H. 1991. "Quantitative Quality Tests for Frozen Fish: Soluble Protein and Free Fatty Acid Content as Quality Criteria for Hake (*Merluccius capensis*) stored at $-18°C$. *J. Sci. Food Agric.* 54:449–458.

KOSTUCH, S. 1978. *Rola Przemian Tlenku Trójmetyloaminy w Zamrażalniczych Zmianach Białek Ryb*. Thesis. Politechnika Gdańska.

KOSTUCH, S., and SIKORSKI, Z. E. 1977. "Interaction of Formaldehyde with Cod Proteins During Frozen Storage." Pp. 1–9. International Institute of Refrigeration. Joint Meeting of Commissions C1 and C2. Proceedings. Freezing, Freeze-Storage and Freeze-Drying of Biological Materials and Food-Stuffs. I.I.F.–I.I.R.–Commissions C1, C2–Karlsruhe.

KRUEGER, D. J., and FENNEMA, O. R. 1989. "Effect of Chemical Additives on Toughening of Fillets of Frozen Alaska Pollack (*Theragra chalcogramma*)." *J. Food Sci.* 54:1101–1106.

LAMPILA, L. E., and BROWN, W. D. 1986. "Changes in the Microstructure of Skipjack Tuna During Frozen Storage and Heat Treatment." *Food Microstructure.* 5:25–31.

LAN, C. C., PAN, B. S., and JIANG, S. T. 1987. "Effect of Reducing Agents on

Protein Denaturation of Frozen Minced Lizard Fish and Textural Properties of the Cooked Product." *J. Chinese Agric. Chem. Soc.* 25:159–168.

LeBLANC, E. L., and LeBLANC, R. J. 1992a. "Determination of Hydrophobicity and Reactive Groups in Proteins of Cod (*Gadus morhua*) Muscle During Frozen Storage." *Food Chem.* 43:3–11.

LeBLANC, E. L., and LeBLANC, R. J. 1992b. "Separation of Cod (*Gadus morhua*) Fillet Proteins by Electrophoresis and HPLC After Various Frozen Storage Treatments." *J. Food Sci.* 54:827–834.

LEBLANC, E. L., LEBLANC, R. J., and GILL, T. A. 1987. "Effects of Pressure Processing on Frozen Stored Muscle Proteins of Atlantic Cod (*Gadus morhua*) Fillets." *J. Food Processing Preservation.* 11:209–235.

LEWIN, S. 1974. *Displacement of Water and its Control of Biochemical Reactions.* London: Academic Press.

LOVE, R. M. 1966. "The Freezing of Animal Tissue." Pp. 317–405. In *Cryobiology,* edited by H. T. Meryman. London/New York: Academic Press.

LUNDSTROM, R. C., CORREIA, F. F., and WILHELM, K. A. 1982. "Enzymatic Dimethylamine and Formaldehyde Production in Minced American Plaice and Blackback Flounder Mixed with a Red Hake TMAO-ase Active Fraction." *J. Food Sci.* 47:1305–1310.

MAO, W. W., and STERLING, C. 1970. "Parameters of Texture Changes in Processed Fish: Cross-linkage of Proteins." *J. Texture Studies* 1:484–490.

MATSUMOTO, J. J. 1979. "Denaturation of Fish Muscle Proteins During Frozen Storage." Pp. 205–224. In *Proteins at Low Temperatures,* edited by Owen Fennema. Advances in Chemistry Series 180. Washington, DC: American Chemical Society.

MATSUMOTO, J. J., and NOGUCHI, S. F. 1992. "Cryostabilization of Protein in Surimi." Pp. 357–388. In *Surimi Technology,* edited by T. C. Lanier and C. M. Lee. New York / Basel / Hong Kong: Marcel Dekker.

NITISEWOJO, P. 1987. "Effect of Frozen Storage on the Texture of Squid (*Loligo* sp.) Mantle." *ASEAN Food J.* 3:72–73.

NOGUCHI, S., and MATSUMOTO, J. J. 1970. "Studies on the Control of the Denaturation of the Fish Muscle Proteins During the Frozen Storage—I. Preventive Effect of Na–Glutamate." *Bull. Japan. Soc. Sci. Fish.* 36:1078–1087.

OWUSU-ANSAH, Y. J., and HULTIN, H. O. 1984. "Trimethylamine Oxide Prevents Insolubilization of Red Hake Muscle Proteins During Frozen Storage." *J. Agric. Food Chem.* 32:1032–1035.

OWUSU-ANSAH, Y. J., and HULTIN, H. O. 1986. "Chemical and Physical Changes in Red Hake Fillets During Frozen Storage." *J. Food Sci.* 51:1402–1406.

OWUSU-ANSAH, Y. J., and HULTIN, H. O. 1987. "Effect of *In Situ* Formaldehyde Production on Solubility and Cross-Linking of Proteins of Minced Red Hake Muscle During Frozen Storage." *J. Food Biochem.* 11:17–19.

OWUSU-ANSAH, Y. J., and HULTIN, H. O. 1992. "Differential Insolubilization of

Red Hake Muscle Proteins During Frozen Storage." *J. Food Sci.* 57:265–266.

PLANK, R., EHRENBAUM, E., and REUTER, K. 1916. *Die Konservierung von Fischen durch das Gefrierverfahren.* Abhandlung Volksernährung, H. 5. Berlin: Verlag Zentral-Einkaufsgesellschaft.

RAGNARSSON, K., and REGENSTEIN, J. M. 1989. "Changes in Electrophoretic Patterns of Gadoid and Non-gadoid Fish Muscle During Frozen Storage." *J. Food Sci.* 54:819–823.

REHBEIN, H., and ORLICK, B. 1990. "Comparison of the Contribution of Formaldehyde and Lipid Oxidation Products to Protein Denaturation and Texture Deterioration During Frozen Storage of Minced Ice-Fish Fillets (*Champsocephalus gunnari* and *Pseudochaenichthys georgianus*)." *Internat. J. Refrig.* 13:336–341.

SANTOS, E. E. M., and REGENSTEIN, J. M. 1990. "Effects of Vacuum Packaging Glazing, and Erythorbic Acid on the Shelf-life of Frozen White Hake and Mackerel." *J. Food Sci.* 55:64–70.

SHENOUDA, S. V. 1980. "Theories of Protein Denaturation During Frozen Storage of Fish Flesh." Pp. 275–311. In *Advances in Food Research,* edited by C. O. Chichester, E. M. Mrak, and G. F. Stewart. Vol. 26. New York: Academic Press.

SIKORSKI, Z. E. 1980. *Technologia Żywności Pochodzenia Morskiego.* Warszawa: Wydawnictwa Naukowo-Techniczne.

SIKORSKI, Z. E., and KOŁAKOWSKA, A. 1990. "Freezing of Marine Food." Pp. 111–124. In *Seafood: Resources, Nutritional Composition, and Preservation,* edited by Z. E. Sikorski. Boca Raton, FL: CRC Press.

SIKORSKI, Z. E., KOŁAKOWSKA, A., and PAN, B. S. 1990. "The Nutritive Composition of the Major Groups of Marine Food Organisms." Pp. 29–54. In *Seafood: Resources, Nutritional Composition, and Preservation,* edited by Z. E. Sikorski. Boca Raton, FL: CRC Press.

SIKORSKI, Z. E., and KOSTUCH, S. 1982. "Trimethylamine N-Oxide Demethylase. Its Occurrence, Properties, and Role in Technological Changes in Frozen Fish." *Food Chem.* 9:213–222.

SIKORSKI, Z. E., KOSTUCH, S., and KOŁODZIEJSKA, I. 1975. "Zur Frage der Proteindenaturierung im gefrorenen Fischfleisch." *Nahrung.* 19:997–1003.

SIKORSKI, Z. E., OLLEY, J., and KOSTUCH, S. 1976. "Protein Changes in Frozen Fish." *Crit. Rev. Food Sci. Nutr.* 8:97–129.

SIROIS, M. E., SLABYJ, B. M., TRUE, R. H., and MARTIN, R. E. 1991. "Effect of Vacuum Packaging on Changes Associated with Frozen Cod Fillets." *J. Muscle Foods* 2:197–208.

SUZUKI, T. 1981. *Fish and Krill Protein: Processing Technology.* London: Applied Science Publishers.

TAKAMA, K., ZAMA, K., and IGARASHI, H. 1972. "Changes in the Flesh Lipids of

Fish During Frozen Storage. Part III. Relation Between Rancidity in Fish Flesh and Protein Extractability." *Bull. Japan. Soc. Sci. Fish.* 38:607–612.

TOKUNAGA, T. 1964. "Studies on the Development of Dimethylamine and Formaldehyde in Alaska Pollack During Frozen Storage. I." *Rep. Hokkaido Reg. Fish. Res. Inst.* 29:108–122.

TRAN, V. D. 1974. "The Preventive Effect of Pyruvic Acid and Other Ketoacids on the Denaturation of Proteins in Frozen Fish." *Can. Inst. Food Sci. Technol. J.* 7:203–208.

YOON, K. S., and LEE, C. M. 1990. "Effect of Powdered Cellulose on the Texture and Freeze–Thaw Stability of Surimi-Based Analog Products." *J. Food Sci.* 55:87–91.

9

CHANGES IN PROTEINS AND NONPROTEIN NITROGEN COMPOUNDS IN CURED, FERMENTED, AND DRIED SEAFOODS

Zdzisław E. Sikorski and Adriaan Ruiter

INTRODUCTION

The objective of salting, fermenting, marinating, drying, and smoking of fish and marine invertebrates is to increase the shelf life and to develop in the products the desirable sensory properties. The storage life extension in various products is achieved mainly due to the different combined effects of low water activity, reduction in the number of vegetative forms of the microflora by heat, the preservative action of acetic acid and added chemical preservatives, and the antibacterial and antioxidant activity of various smoke components. The sensory properties of the products, typical for the different commodities, develop as a result of changes in lipids and nitrogenous compounds in the tissues caused by chemical and enzymatic processes controlled by the parameters of processing, principally the temperature, pH, and the presence of salts and other reactive components (Doe and Olley, 1990; Shenderyuk and Bykowski, 1990; Miler and Sikorski, 1990).

SALTING

Factors Affecting Proteins in Salted Seafoods

In salted seafoods, preserved by mixing with dry salt or by immersion in brine, protein changes result from the action of salt and proteolytic enzymes. Due to the combined effect of sodium chloride, of different brine impurities, and of the enzymes, the sensory properties of the meat of several species of fish may be so altered that the product turns suitable for eating without heating during culinary preparation. Furthermore, different additives used in various procedures for salting, for example, sugar and spices, contribute to the final flavor of the product. The conditions during salting, ripening, and storage must be such as to increase the rate of penetration of salt into the tissues, to promote the development of typical sensory properties of the products, and to retard spoilage reactions. The enzymatic and spoilage processes are controlled mainly by the concentration of salt in the tissues, temperature, pH, and the availability of oxygen. The rate of salt penetration into the muscles, which is critical for the control of spoilage and ripening, may increase up to 100-fold by filleting and skinning the fish.

The Effect of Salt

In salting fish, the salt should not be too finely grained because the fine-grained salt dissolves rapidly in the fish muscle fluids causing a too-fast withdrawal of moisture from the surface tissue. As a consequence, a rapid protein denaturation and coagulation occur, preventing further penetration of the salt into the fish and giving rise to a condition known in the trade as "salt burn." In salted fish, this denaturation always occurs, but normally it proceeds slower than the penetration of salt (Tülsner, 1978). The cause of salt denaturation differs from that of heat denaturation. The Na^+ and Cl^- ions act as counterions toward negatively and positively charged groups, respectively, disturbing the native conformation of the proteins. Denaturation by salt, just like denaturation by heat, results in a decreased extractability of fish muscle proteins. Heavily salted fish loses much water, its texture is tough, and the flavor is much less developed than, for example, in fatty herring salted with a low amount of salt.

Another effect of salt has to do with the ion-exchanging properties of proteins. The Ca^{2+} and Mg^{2+} ions present in the muscles may be replaced by Na^+ from the brine, giving rise to a soft, "mushy" texture of the fillet. As this process is reversible, Ca^{2+} and Mg^{2+} from impure salt may accumulate in the flesh, resulting in a tough product. Chemical reactions leading to the formation or cleavage of covalent bonds in proteins may also be induced by the addition of salts because the proteins may be more accessible for such reactions. Catalytic effects of salt cannot be excluded either, but they result mainly from the effects of impurities present in the brine. Traces of heavy metals, in particular copper, may accelerate the Maillard-type reactions (Shewan, 1955).

Enzymatic Changes

According to Biegler (1960) "Ein mild gesalzener Schinken oder mild gesalzener Kaviar oder mild gesalzener Lachs dürften schon den Göttern gut gemundet haben." The exquisite sensory properties of lightly salted sturgeon, salmon, herring, anchovy, or other fatty fish are characteristic because of the tender texture which is that of a viscoelastic body and the typical flavor, which has the fishy, salty, slightly meaty, cheesy, greasy, and possibly very slightly rancid note. These sensory characteristics result from enzymatic changes in proteins, lipids, and added carbohydrates, from different interactions of the products of these enzymatic reactions, and from the added spices. Lipids play an important part in the development of the sensory properties of salted fish because they contribute directly to the texture of the muscle. Furthermore, the products of lipid hydrolysis and oxidation participate in maturing by interacting with nitrogenous compounds and, if present in excessive amounts, they lead to rancidity. Fish salted by adding about 13 kg salt and 1.5 kg of a mixture of different spices per 100 kg of fish may contain about 0.05–0.06% of terpenoids and phenols. These compounds not only impart distinct flavor notes to the product but also exert a significant antioxidant activity.

Protein hydrolysis is accomplished by muscle proteases, as well as by the enzymes of the fish alimentary tract and of bacteria. The activity of the muscle proteases is very small as compared to that of the enzymes of the gastrointestinal tract. Furthermore, the enzymes from various sources have different pH optima (Fig. 9–1) and are affected differently by sodium chloride. The role of proteolytic enzymes in ripening of salted fish has been investigated by Luijpen (1959) and discussed by Meyer (1965).

According to Kowalewski (1989), in whole Baltic sprats, the highest proteolytic activity against the sprat muscle proteins is in the pH range 3–5. The autolytic activity is proportional to the lipid content in the fish, affected by the seasonal changes in feeding intensity. In salted sprats, the rate of proteolysis depends on the amount of salt. In fish containing 10 and 20% salt, it is 1.2–1.5 and 0.5–0.6 of the rate at 13% salt in the

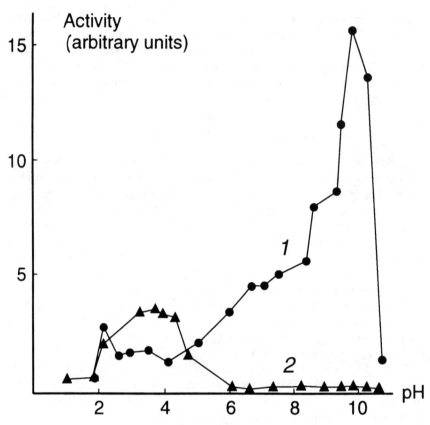

Figure 9–1   Endopeptidase activity of the pyloric caecae (1) and of muscle tissue (2) of maatjes herring as affected by pH.

Source: Data from Luijpen (1959).

muscle, respectively. After the first week of salting, the rate of proteolysis is the highest, being 2.0–3.7 times larger than the initial value. During further ripening and storage of the sprats salted with a mixture of salt and spices, the accumulation of different forms of NPN increases logarithmically with time up to 1 year. The liberation of peptides and amino acids from proteins increases the buffering capacity of the aqueous extract from the salted fish. The change in buffering capacity is linear during the first year and turns into a logarithmic relationship during further storage. The temperature coefficient (Q_{10}) of the ripening reactions in salted Baltic sprats is about 3.1 in the range from 0 to $-10°C$ (Kowalewski, 1989).

The hydrolytic processes bring about significant changes in the texture of the muscles. Prolonged storage of ripe salted sprats, even under refrigeration, leads to excessive softening, which makes the fish unsuitable for filleting and for use in the manufacture of different delicatessen products. Overripe salted sprats may be covered by white deposits of crystallized peptides and amino acids. The first patches of this white "bloom" occur on the surface of sprats at the stage when the buffering capacity value, which at the beginning of the ripening process is 60–90 units, reaches values of 180–210. For overripe sprats, the buffering capacity value is 220–270; for sprats not suitable for sale it is 280–350 (Levieva, 1956). According to Kowalewski (1989), these deposits consist in about 50% of tyrosine, 20% phenylalanine, and 10% glutamic acid.

A fast and pronounced ripening occurs in Dutch "maatjes"-cured herring. This product is made from immature fish caught in or at the end of the feeding period. Such herring is rich in fat and there is a very high proteolytic activity in the gut, especially in the pyloric appendages. The fish is gibbed before salting so that the pancreatic tissue is left in the belly cavity. Thus, the enzymes may act on the muscle proteins during ripening, causing the characteristic texture and flavor due to proteolysis and probably other reactions as well (Luijpen, 1959; Ruiter, 1972b). It should not be excluded that components from the feed also contribute to the flavor of the product, either directly or by participating in different ripening reactions.

Generally, a fish fillet or gutted fish is less suitable for ripening in brine than whole fish. However, it may undergo ripening, just as a fish caught off season of the highest enzyme activity, and of fatty species which normally do not develop the characteristic features of ripened salted products, if suitable proteolytic enzymes are added to the brine. Positive results in salting gutted fish and fillets of Atlantic herring and mackerel on a semicommercial scale were obtained by Góra (1972, 1973). A proteolytic enzyme, Rapidase B–500, was used in amounts of 0.5–0.7 g/kg. At a salt concentration in the tissues of 10%, the ripening of headed and gutted mackerel lasts 1.5–2 months at -1 to $-4°C$ or 1 month at 0–5°C. At a higher salt concentration, the ripening lasts considerably longer.

Typical Scandinavian products such as *kryddersild*, *tidbits*, and *gaffelbiter* are prepared from sugar-salted herring, to which nitrate and some other ingredients are added. The salted fish is kept at low temperatures and has to go through a long ripening process, that is, up to 1.5 years, during which it becomes tender and develops a characteristic flavor. The microbial activity during ripening is generally low and the process is predominantly autolytic (Knøchel and Huss, 1984a). Nitrate, however, is partially reduced by bacteria and the nitrite formed reacts with myoglobin, causing a bright reddish color of the muscle. Knøchel and Huss (1984b) found the level of nitrosodimethylamine in the product to be low.

FERMENTING

Fish Sauces

Fermented fish paste and sauce are popular products prepared and consumed in Southeast Asia as a source of nutrients and as condiments (Beuchat, 1983). Generally, salted fish, often whole, ungutted, is used for preparing such products and protein hydrolysis plays a major part in producing the desirable sensory properties. Typical examples of such products are the *nuoc-mam* produced in Vietnam and Cambodia, *nampla* in Thailand, or *patis* in the Philippines. A fish source resembling *nampla* was prepared from white bait (*Anchoviella* spp.) and carangid (*Caranx caranx*) by mixing the minced whole fish with salt 3:1 and fermenting 12 months at 28–35°C (Hiremath et al., 1985). The white bait and carangid sauces contained total nitrogen 13.3 and 11.6 g/dm^3, amino nitrogen 8.0 and 6.2 g/dm^3, total volatile base nitrogen 0.42 and 0.39 g/dm^3, and volatile acids 2.8 and 3.7 g/dm^3, respectively. The pH of the products were 5.8 and 5.9.

A high-quality fish sauce prepared from overripe, lightly salted, spicy Baltic sprats contains about 1.5% total nitrogen. The NPN and amino acid nitrogen in the product constitute about 85 and 37% of the total nitrogen, respectively. The pH of the sauce is 4.5–5.0 (Kowalewski, 1989). Besides proteolysis decarboxylation of amino acids occurs also in the fermented fish products. Fardiaz and Markakis (1979) identified a number of amines, including histamine and 2-phenylethylamine, in fermented fish pastes.

In the traditional procedure, both autolytic and bacterial processes are involved and the reactions proceed in anaerobic conditions. The microflora of Thai fish sauce consists mainly of halophilic organisms which produce amines and volatile acids (Saisithi et al., 1966). In other fermented fish products from Thailand, *Lactobacillus brevis* and *Pediococcus pentosaceus* were found to be the most important organisms responsible for the typical flavor and texture (Wongkhalaung, 1988).

The quality of fish and shellfish sauces produced by traditional methods may be difficult to control and off-flavors may occur due to the activity of halophilic putrefactive bacteria. Therefore, many attempts were made to produce fermented fish products using modern engineering, microbiology, and enzymology procedures. Herborg and Johansen (1977) prepared a curd called "fish cheese" from minced fish flesh with the aid of a *Lactobacillus* starter culture. A sauce made by adding cultures of *Cellulomonas flavigena* NTOU and its mutants to sterilized homogenized shrimp had a high sensory quality and the yield of nitrogen in the hydrolysate reached 50–63% (Chen et al., 1992). In a study on the effect of bromelain, papain, and ficin, the best results were obtained with bromelain. Addition of this product did not affect the quality of traditional fish sauce (Beddows and Ardeshir, 1979). Although fermentation normally takes place at a high salt content, it may be advantageous to delay salting for some time. In the experiments of Bigueras-Benitez et al. (1988) the rate of proteolysis increased in fish salted 7 h after the addition of bromelain.

Silage

Autolysis occurs also in fish silage prepared for animal feeding (Windsor and Barlow, 1981; Raa and Gildberg, 1982). This method of utilizing fish offal and trash fish is popular in Scandinavian countries. It was first introduced in Finland by Virtanen (Backhoff, 1976) and has been used on a commercial scale in Poland. Recently, ensilaging has been recognized as a very efficient method of utilizing fish offal in developing countries.

In Poland, the minced fish offal is treated with 0.2% sulfuric acid, 0.2% hydrochloric acid, and 2% formic acid to decrease the pH of the pulp to 3.5–3.8 to accelerate autolysis and to inhibit bacterial spoilage. After 1 day of maturation at 40°C, the product takes on a liquid consistency and has a pleasant fishy smell. In a different procedure, which is cheaper, 1% sulfuric and 1.3% sodium metabisulfite is used, which brings the pH to about 4.5. Both products can be stored for months at room temperature and have a high feeding value for hogs and poultry (Sikorski et al., 1969; Bykowski, 1988).

A fully preserved silage of minced cod viscera can be prepared by adding 0.75% (v/w) propionic acid and 0.75% (v/w) formic acid. The product has a pH of 4.3, remains sterile, and retains a fresh acidic smell for at least 1 year at 27°C. After 17 days at this temperature, about 85% of the protein is solubilized and no further solubilization occurs. During storage, there is a slow deamination of amino acid residues in the proteins, corresponding to about 8% of the protein nitrogen for 220 days (Gildberg and Raa, 1977).

The low pH necessary in ensilaging for microbiological safety may be also achieved by mixing the fish offal with proper amounts of different carbohydrate-rich agricultural waste materials or molasses and adding a starter culture of *Lactobacillus plantarum* (Sikorski, 1971; Brown and Sumner, 1985).

The proteolytic processes can be accelerated by adding suitable enzymes, which can be chosen from a large variety of proteinases of animal, plant, or microbial origin, with pH optima ranging from 2 for pepsin to 8.5 for alcalase isolated from *Bacillus* sp. and with a broad range of optimal temperature (Haard et al., 1982; Mackie, 1982). Microbiological putrefaction during the process can be easily prevented by using some of the commercially available plant enzymes, which have an optimum temperature as high as 70°C. The hydrolysis of minced fish/offal in the presence of added enzymes lasts only a few hours at optimum parameters, resulting in solubilizing about 80% of the total nitrogen and about as much of the soluble fraction is composed of peptides and amino acids. The less substrate-specific endopeptidases of plant and bacterial origin are more effective than the animal digestive enzymes. The soluble products can be used as milk replacers for young animals.

The ensilaging as well as hydrolysis of fish offal with the aid of added enzymes is treated in some details in Chapters 12 and 13.

MARINATING

In producing raw marinades, the seafood, mainly herring, is ripened at room temperature in a marinating bath consisting of a water solution of acetic acid, salt, and sugar. Generally, after 1–2 weeks or after longer periods of storage under refrigeration, the product is packaged in suitable containers with a cover brine or sauce containing different spices and condiments. The desirable sensory properties are developed due to the ripening process as well as are imparted by the components of the cover brine. Other marinated assortments are prepared by cooking or frying the seafood, for example, herring, cod, eel, or mussels, and packaging the product with suitable cover brine, often containing gelatine. The sensory properties of these products are developed as a result of cooking or frying and are also affected by the components of the brine or sauce.

The marinating process brings about denaturation of fish proteins by sodium chloride and acid. Denaturation by acetic acid starts at a pH of about 5 and proceeds faster at lower pH values. By binding to the basic amino acid residues, the acetate ion changes the charge of the proteins. Protein hydrolysis induced by acetic acid under the conditions of marinating is rather improbable. If the acid concentration is too high, the skin of the herring may peel off the flesh. This happens also if the product is stored too long.

A pH of 4–4.2, which normally occurs in marinades, is in the optimum range for the activity of several fish muscle proteolytic enzymes. In extreme cases, up to about 50% of the total protein mass in marinated fish undergoes at least partial proteolysis (Tülsner, 1978). Disintegration of the Z-lines in the myofibrils as well as myosin breakdown was demonstrated (Rodger et al., 1984). The protein changes result in a loss of firmness of the muscle. This breakdown, however, is checked by the presence of salt. In some Western European countries, it is common practice to add hydrogen peroxide to the marinating back to bleach the herring. Hydrogen peroxide reacts also with SH groups in proteins which are oxidized to sulfoxide groups and may decrease the proteolytic activity.

In fresh marinated herring, the high acidity prevents the proliferation of most types of spoilage bacteria. However, the heterofermentative *Betabacterium büchneri* and *Betabacterium breve* may decarboxylate amino acids liberated during the ripening process, especially arginine, lysine, histidine, tyrosine, and glutamic acid. The accompanying increase in pH of the medium may cause the development of spoilage microflora.

DRYING

Fish drying is practiced in the open air in a cold or temperate climate in windy terrain of low air humidity as on the North Atlantic coast. In more hot and humid areas solar drying is used. The application of different driers with circulating air of controlled temperature and humidity, as well as freeze drying, makes the process independent of the climate conditions. Seafoods may be dried in whole, if sufficiently small, or are prepared by gutting, splitting, filleting, or even mincing. Many dried-seafood assortments are prepared from cooked or cured raw material, whereas some traditional products are even treated with alkali. Thus, the proteins in the dried tissues are affected by different factors, that is, the rate and extent of loss of water and the accompanying concentration of the mineral components in the tissues, the action of salt, the time and temperature of heat treatment, the autolytic and bacterial proteolytic activity, the pH of the tissues, and the interactions with lipid oxidation products. Depending on the type and extent of changes in proteins and on the development of desirable texture and flavor, the different assortments of dried seafoods may be consumed as they come, for example, the dried Caspian roach (Zaitsev et al., 1969) as a snack compatible with beer and dried abalone as an esoteric appetizer (Olley and Thrower, 1977), or they are used for preparing various dishes, or even as condiments.

The different drying conditions decrease slightly the nutritional properties of seafood proteins, depending on the severity of heat treatment, oxidation, and cross-linking between the amino acid residues. Maillard-type nonenzymatic browning reactions may involve NH_2 groups in amino acid residues, as well as in free amino acids (Pan, 1988). Generally, however, more severe loss is caused by insect infestation in unsuitably packaged dried products during storage at abuse temperature and air humidity (Olley et al., 1988).

SMOKING

There are basically three types of smoked products: (1) cold smoked at temperature not exceeding 30°C, (2) hot smoked in temperatures high enough to cause thermal denaturation of the muscle proteins, and (3) smoke-dried in which the fish is treated as in (2) and subsequently dried. For heavily smoked and dried products, the primary objective of processing is extension of the shelf life, whereas in soft-cured products, the delicious sensory properties are more important.

In smoked fish, the properties of proteins are affected in ways similar to those in dried products. One additional factor is the chemical reactivity of a large number of components present in the wood smoke (Kurko, 1969). These components are not only responsible for the smoky flavor of the products but they also have very significant antioxidant activity. They react with many functional groups in protein–amino acid residues. In model systems, the carbonyl and phenol fractions of wood smoke decrease the contents of SH groups in cysteine, glutathione, and meat (Krylova et al., 1962). Treatment of proteinaceous foodstuffs, proteins, and amino acids with wood smoke brings about a substantial reduction of the contents of amino groups (Krylova et al., 1962) and of available lysine (Dvořák and Vognarova, 1965; Chen and Issenberg, 1972).

Some of the reactions of smoke components with proteins and NPN compounds contribute to the formation of the desirable golden color of smoked fish. The importance of nonenzymic browning reactions in smoke color formation was already pointed out by Ziemba (1967). It was demonstrated that glycolic aldehyde and methyl glyoxal, both present in considerable amounts in wood smoke, are active browners with amino groups and that formaldehyde is also incorporated in the structures formed (Ruiter, 1970, 1972a, 1979). In an experiment with collagen sausage casings, a strong decrease in available lysine was observed while arginine remained unchanged (Ruiter, 1970). When, in another experiment, a collagen casing was used in which the NH_2 groups were removed, no color developed during smoking. These experiments confirm the key role of the amino groups in smoke color formation.

The surface of smoked fish is covered by a glossy, bright, shining pellicle. According to Shewan (1944), this pellicle results from drying the superficial sticky layer of protein peptized by the action of the brine and probably due also to deposition on the surface of resinous condensation products from the smoke. A glossy, sticky, brown pellicle can be deposited on the smoked surface, even a protein-free metal sheet, in just 3 min of electrostatic smoking (Sikorski, 1971).

Regardless of the reactivity of smoke components toward the functional groups in protein–amino acid residues, there is no substantial decrease in the nutritional value of smoked fish, as most of the products are rather lightly smoked and most of the smoke components hardly penetrate deeper than 1 mm under the skin of the fish.

REFERENCES

BACKHOFF, H. P. 1976. "Some Chemical Changes in Fish Silage." *J. Food Technol.* 11:353–363.

BEDDOWS, C. G., and ARDESHIR, A. G. 1979. "The Production of Soluble Fish Protein Solution for Use in Fish Sauce Manufacture. I. The Use of Added Enzymes." *J. Food Technol.* 14:603–612.

BEUCHAT, L. R. 1983. "Indigenous Fermented Foods." Pp. 477–528. In *Biotechnology*, edited by H.-J. Rehm and G. Reed. Vol. 5. Weinheim: Verlag Chemie.

BIEGLER, P. 1960. "Fischwaren-Technologie. Theorie und Praxis der Fabrikationsmethoden zur Konservierung von Fischen." In *Der Fisch*, edited by C. Baader. Vol. 5. Lübeck, Germany: Verlag Chemie.

BIGUERAS-BENITEZ, C. M., GUZMAN, D. L., and ANCHETA, E. C. 1988. "Delay in Salting and Use of Proteolytic Enzyme in the Manufacture of Fish Sauce from Round Scad (*Decapterus makrosoma* Bleeker)." *Fish. Tech. News (FAO)*, No. 11, 6.

BROWN, N., and SUMNER, J. 1985. "Fish Silage." Pp. 404–414. In *Spoilage of Tropical Fish and Product Development*, edited by A. Railly, FAO Fisheries Report No. 317 Supplement. Rome: Food and Agriculture Organization.

BYKOWSKI, P. 1988. "Kierunki i Możliwości Rozwoju Produkcji Hydrolizatów Paszowych z Ryb." *Bull. Sea Fish. Inst. Gdynia* 19(3–4):3–12.

CHEN, L. B., and ISSENBERG, P. 1972. "Interactions of Wood Smoke Components with ε-Amino Groups in Proteins." *J. Agric. Food Chem.* 20:1113–1115.

CHEN, H. CH., HO, W. L., MOODY, M. W., JIANG, S. T. 1992. Modification of *Cellulomonas flavigena* NTOU 1 characteristics for the production of shrimp hydrolysates. *J. Food Sci.* 57:271–276.

DOE, P., and OLLEY, J. 1990. "Drying and Dried Fish Products." Pp. 125–145. In *Seafood: Resources, Nutritional Composition, and Preservation,* edited by Z. E. Sikorski. Boca Raton, FL: CRC Press.

DVORAK, Z., and VOGNAROVA, I. 1965. "Available Lysine in Meat and Meat Products." *J. Sci. Food Agric.* 16:305–312.

FARDIAZ, D., and MARKAKIS, P. 1979. "Amines in Fermented Fish Paste." *J. Food Sci.* 44:1562–1563.

GILDBERG, A., and RAA, J. 1977. "Properties of a Propionic Acid/Formic Acid Preserved Silage of Cod Viscera." *J. Sci. Food Agric.* 27:647–653.

GORA, A. 1972. "Opracowanie Technologii Słabego Solenia Śledzi i Makreli oraz Enzymatycznego Solenia Filetów Śledziowych." *Research Report No. 203.3(T–95).* Sea Fisheries Institute, Gdynia.

GORA, A. 1973. "Próby Enzymatycznego Solenia Makreli w Skali Półtechnicznej." *Research Report No. 203(T–99/I).* Sea Fisheries Institute, Gdynia.

HAARD, N. F., FELTHAM, L. A. W., HELBIG, N., and SQUIRES, E. J. 1982. "Modification of Proteins with Proteolytic Enzymes from the Marine Environment." Pp. 223–244. In *Modification of Proteins. Food, Nutritional, and Pharmacological Aspects,* edited by R. E. Feeney and J. R. Whitaker. Advances in Chemistry Series 198. Washington, DC: American Chemical Society.

HERBORG, L., and JOHANSEN, S. 1977. "Fish Cheese: The Preservation of Minced Fish by Fermentation." Pp. 353–355. *Proc. Conf. on the Handling, Processing and Marketing Tropical Fish.* London: Tropical Products Institute.

HIREMATH, G. G., BHANDARY, M. H., HANUMANTHAPPA, B., SUDHAKAR, N. S., and SHETTY, H. P. C. 1985. "Production of Fish Sauce from Two Species in Indian Fishes." Pp. 393–402. In *Spoilage of Tropical Fish and Product Development,* edited by A. Railly, FAO Fisheries Rep. No. 317 Supplement. Rome: Food and Agriculture Organization.

KNØCHEL, S., and HUSS, H. H. 1984a. "Ripening and Spoilage of Sugar Salted Herring with and Without Nitrate. I. Microbiological and Related Chemical Changes." *J. Food Technol.* 19:203–213.

KNØCHEL, S., and HUSS, H. H. 1984b. "Ripening and Spoilage of Sugar Salted Herring with and Without Nitrate. II. Effect of Nitrate." *J. Food Technol.* 19:215–224.

KOWALEWSKI, W. 1989. *Sterowanie Procesami Proteolitycznymi w Dojrzewaniu Ryb Solonych—na Przykładzie Szprota Bałtyckiego.* Ph.D. thesis. Politechnika Gdańska.

KRYLOVA, N. N., BASAROVA, K. I., and KUZNETSOVA, V. V. 1962. "Interaction of

Smoke Components with Meat Constituents." Paper presented at the 8th European Meeting of Meat Research Institutes. Moscow.

KURKO, V. I. 1969. *Khimya kopchenya.* Moskva: Pischchevaya Promyshlennost.

LEVIEVA, L. C. 1956. "Bufernost kak Obeektivnyi Pokazatel Stepeni Zrelosti Prezervov." *Rybnoe Khozyaistvo.* 32(5):81–83.

LUIJPEN, A. F. M. G. 1959. *De Invloed van het Kaken op de Rijping van Gezouten Maatjesharing.* Utrecht: Diss., 91 pp.

MACKIE, I. M. 1982. "Fish Protein Hydrolysates." *Process Biochem.* 17:26–28, 31.

MEYER, V. 1965. "Uber die Bedeutung proteolytischer Fermente bei der Herstellung nichtsterilisierter Fischwaren." *Arch. Fischereiwissensch.* 15:245–253.

MILER, K. M. B., and SIKORSKI, Z. E. 1990. "Smoking." Pp. 163–180. In *Seafood: Resources, Nutritional Composition, and Preservation,* edited by Z. E. Sikorski. Boca Raton, FL: CRC Press.

OLLEY, J., and THROWER, S. J. 1977. Abalone—An Esoteric Food. Pp. 143–186. In *Advances in Food Research,* edited by C. O. Chichester, E. M. Mrak, and G. F. Stewart, Vol. 3. New York: Academic Press.

OLLEY, J., DOE, P. E., and HERUWATI, E. S. 1988. "The Influence of Drying and Smoking on the Nutritional Properties of Fish: An Introductory Overview." Pp. 1–21. In *Fish Smoking and Drying. The Effect of Smoking and Drying on the Nutritional Properties of Fish,* edited by J. R. Burt. London/New York: Elsevier Applied Science.

PAN, B. S. 1988. "Undesirable Factors in Dried Fish Products." Pp. 61–71. In *Fish Smoking and Drying. The Effect of Smoking and Drying on the Nutritional Properties of Fish,* edited by J. R. Burt. London/New York: Elsevier Applied Science.

RAA, J., and GILDBERG, A. 1982. "Fish Silage—A Review." *Crit. Rev. Food Sci. Nutr.* 18:383–419.

RODGER, G., HASTINGS, R., CRYNE, C., and BAILEY, J. 1984. "Diffusion Properties of Salt and Acetic Acid into Herring and Their Subsequent Effect on the Muscle Tissue." *J. Food Sci.* 49:714–720.

RUITER, A. 1970. "The Colouring of Protein Surfaces by the Action of Smoke." *Lebensm. Wiss. Technol.* 3:98–100.

RUITER, A. 1972a. "The Browning Reaction of Glycolic Aldehyde with Aminoethanol. II. Effect of Formaldehyde." *Lebensm. Wiss. Technol.* 5:137–139.

RUITER, A. 1972b. "Substitution of Proteases in the Enzymic Ripening of Herring." *Ann. Technol. Agric.* 21:597–605.

RUITER, A. 1979. "Color of Smoked Foods." *Food Technol.* 33(5):54, 56, 58–60, 63.

SAISITHI, P., KASEMSARN, B. O., LISTON, J., and DOLLAR, A. M. 1966. "Microbiology and Chemistry of Fermented Fish." *J. Food Sci.* 31:105–110.

SHENDERYUK, V. I., and BYKOWSKI, P. J. "Salting and Marinating of Fish." Pp. 147–162. In *Seafood: Resources, Nutritional Composition, and Preservation,* edited by Z. E. Sikorski. Boca Raton, FL: CRC Press.

SHEWAN, J. M. 1944. "The Effect of Smoke Curing and Salt Curing on the Composition, Keeping Quality and Culinary Properties of Fish." *Proc. Nutr. Soc.* 2(1–2):105–112.

SHEWAN, J. M. 1955. "The Browning of Salt-Cured White Fish." *Food Manufacture* 30:200–203.

SIKORSKI, Z. E. 1971. *Technologia Żywności Pochodzenia Morskiego.* Warszawa: Wydawnictwa Naukowo-Techniczne.

SIKORSKI, Z. E., DUNAJSKI, E., and KOBYLIŃSKI, S. 1969. "The Reduction of Fish in Poland. *FAO Fisheries Technical Paper* 69, p. 17.

TÜLSNER, M. 1978. "Die bestimmten stofflichen Veränderungen in Fisch beim Salzen und Marinieren." *Lebensmittelindustrie.* 25(4):169–73.

WINDSOR, M., and BARLOW, S. 1981. *Introduction to Fishery By-Products.* Farnham, Surrey: Fishing News Books.

WONGKHALAUNG, C. 1988. "Improvement of a Thai Traditional Fermented Product." Seventh Session of the IPFC Working Party on Fish Technology and Marketing, Bangkok, Thailand April 19–22, 1988. Abstracts of Papers. *Fish Tech. News (FAO)*, No. 11, page 1.

ZAITSEV, V., KIZEVETTER, L., MAKAROVA, T., MINDER, L., and PODSEVALOV, V. 1969. *Fish Curing and Processing.* Moscow: MIR Publishers.

ZIEMBA, Z. 1967. "Nonenzymatic Browning Reactions in Smoked Foods." *J. Nutr. Dietet.* 4:122–129.

10

Functional Food Protein Ingredients from Fish

Tyre C. Lanier

INTRODUCTION

It is increasingly common to encounter commercially processed foods whose constituent carbohydrates, lipids, and proteins are derived from several differing natural sources. Often these food-ingredient materials have been refined as semipure fractions before being used as food ingredients. They may even represent the product of complex chemical and/or physical alterations to a naturally derived material, which, like cellulose for example, may not normally be a source of human food. Such refined and/or altered substances constitute the bulk of the present-day food-ingredient market, a vast assortment valued solely for, and specified by, their individual "functional properties." This term is applicable only to materials destined to be food ingredients, referring to their individual ability to contribute to the taste, texture, and appearance of a formulated food.

We recognize why formulated foods are gaining in market share over the more straightforward foodstuffs: They offer processors control over product quality, portion size, and costs, also reducing waste and improving machinability. More importantly, consumers generally prefer the consistent quality ("no surprises") attainable with formulated foods over the high variability of natural foodstuffs, even when the latter occasionally provides more exceptional quality. Often only a formulated food can offer the nutritional features (less fat, cholesterol, etc.) desired by consumers in familiar product forms. Formulated foods also meet the basic human craving for novelty by broadening the number of possible product forms.

What are the implications of this paradigm for the marketing of fish? Presently, there is still a tremendous wastage in the use of fish protein as food. This situation has arisen primarily from a decided preference for specific species, and portions of these species, to produce the rather limited array of fish products customarily consumed. The technologies of minced fish and surimi production have offered some hope for more efficient utilization of both utilized and underutilized species of fish. Yet, a food-ingredient mentality is needed to realize the full benefits of these technologies, not the commodity approach of the past.

We should consider fish not merely as a food—that is, muscle tissue processed into traditional forms with well-recognized taste, texture, and appearance traits—but, rather, as a source of functional food ingredients. Many agricultural commodities, such as corn, wheat, potatoes, and so on, are commonly fractionated into components of various purity and differing functional properties for the food-ingredient market. A particularly striking example of this is the soybean, which in Western cultures is almost never consumed in whole form but whose protein and oil components are found in literally hundreds of food items. By analogy, many fatty fish species could become the "soybeans of the sea" (Lanier, 1985a), valued mainly for the functional qualities of their refined oils and proteins.

Mackie was perhaps the first to explore the greater possibilities of fish proteins as food ingredients in his excellent review of a decade ago (Mackie, 1983). Since that time, despite the development of a "back to nature" mentality with regard to food preferences among consumers in Western cultures, acceptance of formulated foods has continued to increase. In particular, an amazing acceptance of surimi-based foods by Western peoples has developed, This, and a greater acceptance in the diversity of ingredient sources being used in formulated foods, are trends that bode well for increased consumer acceptance of foods containing ingredients derived from fish.

Nutritional benefits not withstanding, the value of a food ingredient consists primarily in the uniqueness of the functional properties it exhibits in foods. The grand concept of utilizing the world's ocean resources for food in the forms of fish protein concentrates and hydrolysates has consistently met with failure because this fact was overlooked. Perhaps it is true that the world's hungry do need more protein, but experience has proven that it must be delivered in forms that fall within the organoleptic limits established by the traditional eating habits of a culture (Pariser et al., 1978). So if we are to convert fish into protein food ingredients, success will be more likely when these ingredients have a functional value in the food systems already familiar to a populace.

This review will highlight areas of particular present opportunity, and technologies which may open further opportunities, for fish proteins as food ingredients.

FISH PROTEIN HYDROLYSATES AND CONCENTRATES: POTENTIAL FOR ENHANCED FUNCTIONALITY

Fish hydrolysates have for centuries been appreciated in the East as flavoring agents, particularly in the form of fish sauce. This product is now coming into some usage in Western cultures as Eastern cooking techniques have become more popular. Other types of desirable flavorings are now being produced by protein hydrolysis of fish (Haard, 1992; In, 1990). Outside of their value as flavorants, however, hydrolysates compete poorly with other proteins as, without chemical modification, they generally display poor functional properties (Mackie, 1983). Generally also, proteins which are valued for these properties are preferred to be bland and colorless so as to be useful in the broadest range of foods. With few exceptions, hydrolysates are neither, generally exhibiting fishy and bitter flavor notes and a brown color.

Fish protein concentrates (FPC) may be prepared to be relatively bland and light in color but generally exhibit poor solubility and other functional properties (Mackie, 1983; Holden, 1971). Even when solubilized by alkali treatment (Tannenbaum et al., 1970a, 1970b) the proteins are too highly denatured to gel, although reasonable foaming (but not foam stability) properties have been reported (Baldwin and Sinthvalai, 1974).

Much of the effort devoted to chemical modification of FPC or fish protein hydrolysates (FPH) has been for the purpose of restoring or enhancing desired functional properties. Both succinylation and acetylation of partially hydrolyzed fish proteins have been shown to enhance functional properties, particularly the foaming and emulsifying ability, of fish proteins (Spinelli et al., 1977; Brekke and Eisele, 1981). Although some successes have been realized, questions of safety and loss in nutritional value still cloud the future of such chemically modified proteins as food ingredients (Mackie, 1983; Pomeranz, 1992; Mauron, 1979). Recent successes in the biological modification of plant proteins (i.e., by gene insertion into the DNA of the plant) indicate that this approach may offer the ability to more precisely and acceptably modify the functional properties of proteins for food uses (Utsumi and Kito, 1991). It is no doubt a simpler effort to accomplish such modifications in plants than in animals and also likely that genetic manipulation of animals will be subject to greater regulation (Hallerman et al., 1990; Creamer et al., 1988).

For the reason given, commercial efforts at marketing FPC and FPH as foods or food ingredients have declined, and these materials are more seriously considered today for feed or fertilizer (Mackie, 1983; Pariser et al., 1978; Goldhor and Regenstein, 1988). Yet there remains a glimmer of hope on the horizon for certain specialized forms of FPC and FPH which may offer specific functional advantages.

Creamy Fish Protein

A smooth paste can be produced from fish hydrolyzed under controlled conditions and heated to arrest hydrolysis at the proper point. Although the viscera, skins, and fins of the fish are removed prior to processing, the bones remain and are hydrolyzed along with the muscle. The product of this reaction has been termed "creamy fish protein" (CFP) and is being commercially offered as a new fish-derived protein ingredient for foods in Japan. When made from Alaska pollock, the color is very light, and the product is said to possess no bitter flavors (Shoji, 1990).

To date, however, no specific food applications have been demonstrated in which CFP has been shown to excel as an ingredient. Functionality appears to be lacking and, as with FPC, the manufacturers are primarily promoting its nutritional properties. Possibly, however, its relative blandness, light color, and unique texture may yet prove to be desirable properties in new food applications.

Marinbeef—Texturized FPC

The Marinbeef® (*Niigata Engineering Co. Ltd., Tokyo*) process was developed to yield a protein product from fish not dissimilar to the textured vegetable proteins derived from soy (Mackie, 1983; Suzuki, 1981). The process resembles that for FPC in that a solvent extraction is used to remove lipids. Its primary applications were to be in food systems wherein it would substitute for minced meat or poultry, following rehydration from the dried form. There has been little demand for the material in the more developed Western countries, however, due to cost competition from similar soy-derived products, ample supplies of relatively economical beef and poultry minces, and a general cultural bias against fish-derived ingredients in traditional meat and poultry dishes. Success has been greater in countries where muscle protein is at a premium and fish is more commonly eaten (Niki et al., 1992).

Although conversion into Marinbeef totally denatures the fish muscle protein, the process, nonetheless, delivers a material with texturizing and water-sorption properties of functional value to food processors. A further development along these lines is fiberized fish protein, wherein musclelike fibers are spun from a fish protein dope (Mackie, 1983; Horisberger 1979; Tanaka et al. 1983). Again, the application is nearly identical to spun soy fibers (Seal 1977), which also, unfortunately, are able to outcompete fish-derived fibers in the food-ingredient market on the basis of cost. Soy-derived ingredients seem also to be esthetically more appealing to Western consumers in the meat/poultry applications most logical for this market.

Although large markets for these fiberized/denatured protein forms have not as yet developed, the technologies should not be overlooked as holding potential for producing marketable forms of fish proteins. Acceptance of seafoods in general and fish in particular is increasing among Western consumers. New sources of raw materials for such products, such as trimmings from the expanding fillet markets for traditional and newly introduced species, are becoming available. Fabricated foods will continue to gain in acceptance, and fish-based ingredients are in line with the trends in healthy eating and the developing taste for Eastern dishes in Western markets. Certainly, fibrosity (meaty texture) is a marketable functional property for many food applications.

Plasteins—Texturized FPH

Perhaps FPH may also be utilized in texturized forms in the future through utilization of the plastein reaction. Formed by reaction of FPH with certain proteinases under controlled conditions, plasteins are peptide aggregates, associated mainly by hydrophobic interactions (Arai and Fujimaki, 1978), which are insoluble and can possess gellike properties. Furthermore, they do not exhibit many of the bitter reversion flavors that FPH can possess. Conceivably, plasteins might be texturized in some way to be more meatlike, rendering them more functional as food ingredients. Plasteins, being gellike, could also function as fillers of other food gels, an application for which only surimi gels are presently being considered as will be discussed below.

MINCED FISH AS A FOOD INGREDIENT

The production of minced, or mechanically deboned, fish represents the primary step in isolating the proteins of fish for food-ingredient use, that of separating muscle from scale, bone, skin, and viscera. Its production is also the initial step in surimi manufacture (Fig. 10–1). Fish minces as food ingredients have the functional advantage of exhibiting a more meatlike texture than surimi or even mechanically deboned mammalian and poultry meats. In a typical fish deboner, a thick rubber belt presses the fish against a perforated drum, forcing the flesh into the interior of the drum and leaving bones and skin on the exterior to be scraped away (Fig. 10–2). The perforations are typically 3–5 mm, ensuring a coarser particle than that produced in the fine-screen, screwpress-type deboners commonly used for poultry and meats. Surimi also has a finer texture because of a refining step in which leached mince is passed through a small-orifice (about 1 mm) screen to remove any residual scale or bone particles. It follows that, barring the use of some retexturization process for surimi, mince is more suited in texture for use in formed fish portion and stick products. The texture of surimi is irrelevant to its intended use as a gelling agent.

Minced muscle from very lean fish species, obtained by proper processing techniques, is often only slightly less functional in terms of gelling ability than surimi when freshly made (MacDonald et al., 1990a; Hastings, 1989), but its stability in frozen storage is much poorer due to the higher content of active enzyme systems and their substrates (Suzuki, 1981). Minces from fatty species are highly unstable in storage, being more subject to spoilage by lipid oxidation (Hultin, 1987) which additionally contributes to protein denaturation and loss of gelling properties (Shenouda, 1980). Some success has been realized in stabilizing the lipids in leached minces of mackeral, a fatty, dark-fleshed species, by the addition of an antioxidant system comprised of sodium ascorbate, sodium tripolyphosphate, and propyl gallate (Kelleher et al., 1992).

A partially leached mince has been produced by catfish processors in Mississippi (USA) by soaking eviscerated fish frames in ice water prior to mixing (Brooks, 1989). A new process specifically designed to produce leached minces utilizes an inclined screw countercurrent washing machine for continuous production (Anonymous, 1988). Leached minces could offer greater stability in cold storage (Suzuki, 1981), besides being lighter and blander in flavor. Mince, leached or unleached, might also be stabilized in cold storage by the addition of cryoprotective additives such as are commonly incorporated in the final step of manufacturing surimi (Lanier et al., 1992; MacDonald et al., 1990b), just prior to freezing

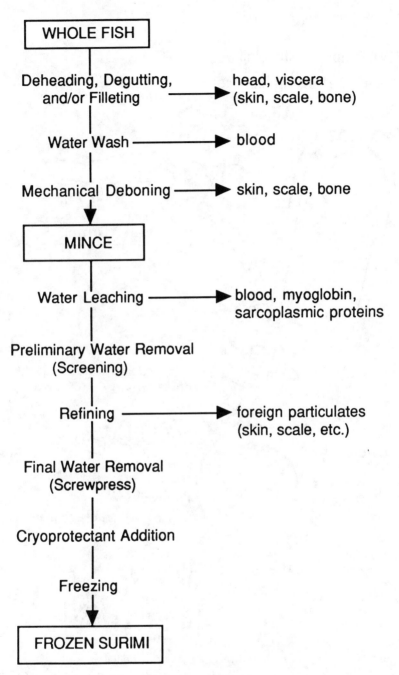

Figure 10–1 The traditional surimi manufacturing process.

Source: Lanier (1986) by permission.

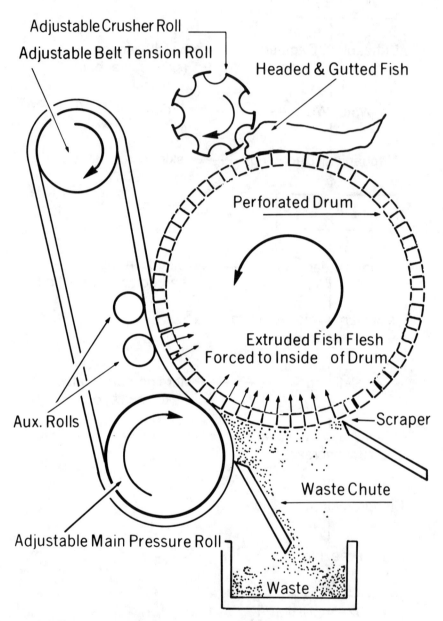

Figure 10–2 Typical design of fish mechanical deboner.

(Fig. 10–1). Cryoprotection of minced New Zealand hoki (Lanier et al., 1992) or Pacific whiting (Kolbe et al., 1992) with sucrose (10–12%) and storage at −20°C has been shown to be equally or more effective in stabilizing the quality than ultralow temperature storage alone. Interestingly, the rate of dimethylamine formation resulting from degradation of trimethylamine oxide, a reaction forming the denaturant formaldehyde as a co-product, was reduced by nearly 50% in cryoprotected hoki mince as compared to an uncryoprotected control (MacDonald et al., 1990b).

Manufacturing a stabilized, unleached mince at sea has been proposed as the first step in a more efficient scheme of producing surimi from deepwater species (MacDonald et al., 1990b). By manufacturing only a stabilized mince, the throughput of factory vessels at sea could be maximized without compromising the quality of the surimi, which would subsequently be produced shoreside where water and labor are less costly to obtain. Alternatively, stabilized minces can offer a more succulent and tender ingredient for traditional fish stick and portion production if the cryoprotectant used is nonsweet and, thus, essentially invisible, such as maltodextrin (MacDonald and Lanier, 1991; Park et al., 1988).

Dried minces have also been developed as food ingredients. One type is currently produced in Norway (Okland, 1986) but is nonfunctional other than serving as a flavor base.

Several international conferences were held in the 1970s to explore potential uses for mechanical deboning technology as applied to fish (Martin, 1972; Martin, 1974; Keay, 1976; Martin, 1980). A chronological review of these proceedings reveals a developing consensus among scientists and industry leaders that the refinement of mince into surimi is a necessary step to produce a more functional, stable, and marketable food ingredient from mechanically recovered fish muscle.

SURIMI—THE FUNCTIONAL FISH PROTEIN CONCENTRATE

Principles of Manufacturing

The surimi manufacturing process takes the refinement process one step further than FPC or minces, effecting not just isolation of the protein, but also a fractionation of the myofibrillar component from the other protein solubility groups (Fig. 10–1). Most of the skin is eliminated by prior skinning or mechanical deboning, and historically the process has been applied mainly to lean fishes so as to avoid problems of fat removal. A concerted research effort sponsored by the Japanese government resulted in several suggested approaches for the physical separation of fat and oil from fattier species to yield a high-protein product. These include removal of the subcutaneous fat layer prior to mechanical deboning by either deep skinning or high-pressure water spray, the latter being accomplished with skinned fillets, water flotation, or centrifugal separation of the fats (Suzuki, 1981), and a novel vacuum leaching process that accelerates the flotation of lipids and high-lipid-content dark meat (Fig. 10–3; Nishioka et al., 1990). Presently, it would appear that vacuum washing holds the most promise as it can remove fats more completely, rapidly, and consistently than the other methods proposed. Fat removal by vacuum washing is most effective when the fish particles being leached are much finer than in a conventional process. To achieve this fine particle size, the fish mince is "micronized" in equipment specially designed to prevent the incorporation of air (oxygen) which might stimulate lipid oxidation (Shimizu et al., 1992).

Whether the raw material is lean or fatty fish, the desired result in surimi processing is a light-colored, preferably white, and bland myofibrillar protein concentrate, low in fat ($< 1.0\%$), connective tissue, and sarcoplasmic protein content, and extremely functional due to the unique gelling properties of fish myofibrillar proteins. This high functionality further distinguishes surimi as being a generation removed from FPC, which is essentially nonfunctional. While FPC was being developed almost simultaneous to the Japanese frozen surimi industry, the worldwide demand for surimi as a food ingredient has soared while demand for FPC has essentially ceased (Mackie, 1983; Gwinn, 1992). Surimi, because

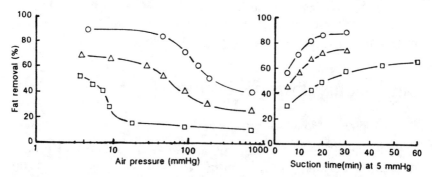

Figure 10–3   Relationship between percent fat removal and air pressure (vacuum; mm Hg) depending on particle size: $\circ$ = 1 mm, $\triangle$ = 3 mm; $\square$ = 5 mm.

Source: Nishioka et al. (1990).

of its unique and superior functional properties, has proven to be the preferred vehicle for delivering fish protein from underutilized species in a form usable by nearly every country and culture. It is certainly the "fish protein concentrate" holding the most promise for the future.

Several innovations to the surimi manufacturing process have been introduced over the past decade which have enhanced the efficiency of the process and led to some improvements in product quality. Decanter centrifuges are now commonly used to recover fine particles lost through the dewatering screens and screwpress (Swafford et al., 1985; Lanier et al., 1992a). A scraped, perforated drum screening device is also now being examined as a lower-cost alternative to centrifuging for fines recovery (Lanier, 1992b). The perforations in the separation screen of this device are cone shaped and extremely small at the scraped surface to prevent clogging but retain very small fines. In-line leaching in tubes or hoses, aided by the placement of static mixing devices at intervals within the pipeline, has presented a means of converting the leaching step to a continuous process without the former problems of low-flow areas where mince might be retained and swell to make subsequent dewatering difficult (Fig. 10–4; Lanier et al., 1992). Drum freezing of surimi has offered the prospect of more rapid freezing (leading to enhanced quality), a lower space requirement, and surimi in a more convenient product form, that of frozen, free-flowing chips (Lanier et al., 1992).

The Unique Gelling Properties of Surimi

Surimi is valued primarily for its gelling property because with the exception of its use in a few products in Eastern cultures, the flavor and color are preferred to be bland so as to broaden the range of acceptable food applications. Upon solubilization with salt and subsequent heating, surimi is converted to a gel. This gel is responsible for the characteristic texture of Japanese kamaboko and similar Eastern fish gel products, and the recently popular shellfish analogs in the West.

Importantly, the gelling properties of surimi are unique. First, gels are firm and more cohesive than those attainable from most other food proteins or hydrocolloids (Lanier, 1986). This cohesive quality is attributed primarily to the quantity and quality of myosin in surimi (Nishioka et al., 1990). Surimi prepared from invertebrate muscle, such as squid and shellfishes, is said to produce particularly cohesive gels, a property attributable to the contribution of paramyosin (Sano et al., 1989), a myosin analog exclusive to such species. The addition of small amounts of invertebrate paramyosin to actomyosin obtained from vertebrate fishes (such as Alaska pollock) has also been shown to increase cohesiveness of the heat-gelled protein (Sano et al., 1986).

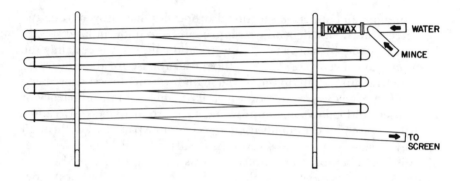

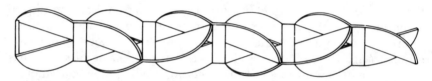

Figure 10–4 Illustration of in-line leaching system, with close-up of komax motionless mixer element.

Source: Holmes (1987) by permission.

Surimi from cold-water fishes also has the unique ability to irreversably gel at very low temperatures. For example, the myofibrillar proteins of homeotherms, such as beef and poultry, do not gel below about 50°C, whereas nonmuscle animal or vegetable proteins require higher temperatures (Montejano et al., 1984). Alaska pollock surimi pastes will, however, form very cohesive gels at 0°C if left several hours undisturbed (Suzuki, 1981).

The Low-Temperature Setting Properties of Surimi

Because of its uniqueness as a functional property among protein ingredients, the cold setting ability of surimi has been intensively studied of late (Niwa, 1992). Modifications to stabilize or enhance this property may be possibly attained through understanding its mechanism. It now appears that cold setting is effected by an inherent transglutaminase enzyme (Kishi et al., 1991; Kimura et al., 1991). This catalyzes a calcium-dependent acyl-transfer reaction in which the γ-carboxamide group of a peptide-bound glutamine residue is the acyl donor, and usually a nucleophilic reagent such as the ε-amino group of lysine or water represents the acyl acceptor, with the formation of a new ε-(γ-glutamyl) lysyl isopeptide bond and concomitant release of ammonia (Fig. 10–5). This creates a covalently bonded gel structure which is quite elastic and cohesive (Lanier, 1986). In the case of pollock, the setting reaction proceeds most completely at 25°C in 2–4 h, although nearly as completely after about 24 h holding at 0°C (Kamath et al., 1992).

Surimi from fish inhabiting warmer waters optimally sets at higher temperatures and may not exhibit setting at all during refrigerated holding (as does Alaska pollock surimi) (Kamath et al., 1992). This suggests that either the transglutaminase is optimally active at different temperatures in different fishes or that the myofibrillar proteins are variably available to and reactive with the enzyme dependent on temperature. On the latter point, several workers (Hashimoto and Arai, 1984; Johnston and Goldspink, 1975) have shown that fish myofibrillar proteins do vary among species in the extent of denaturation exhibited at a given temperature, dependent largely on the environmental growth temperature of the species. This is reflected in a decreased entropy and enthalpy of activation for the inactivation of myofibrillar Ca^{2+}–ATPase in the muscle of species inhabiting colder waters (Fig. 10–6).

Because fish transglutaminase is a calcium-dependent enzyme, its activity can be regulated by the calcium ion content (Saeki et al., 1989). Chelators such as EDTA or citrate are often added to inhibit setting when it may result in clogging of manufacturing equipment during a prolonged holding period. Alternatively, the activity of transglutaminase in surimi could be boosted by the addition of transglutaminase from other sources (Greenberg et al., 1991), such as beef plasma (Kurth and Rogers, 1984) or a recently developed microbial source, *Streptoverticillium* species (Knight, 1990), the latter enzyme not being calcium dependent.

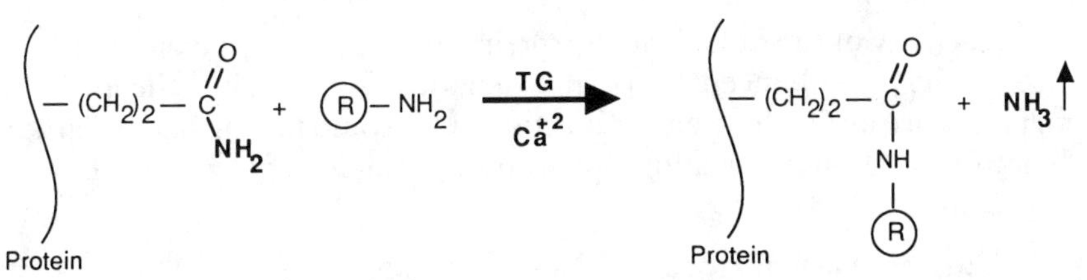

Figure 10–5 Calcium-dependent acyl-transfer reaction catalyzed by transglutaminase.

Source: Greenberg et al. (1991).

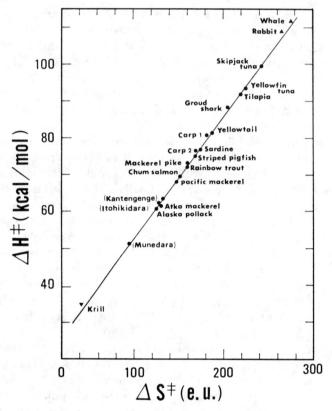

Figure 10–6 Enthalpy ($\Delta H^{\neq}$) and entropy ($\Delta S^{\neq}$) of activation for the inactivation of myofibrillar Ca^{2+}–ATPase of various fish species living at different environmental temperatures.

Source: Hashimoto et al. (1982) by permission.

The reactivity of the myofibrillar protein substrate at any given temperature might be enhanced by partial denaturation, making the protein surface more available to enzymic action. This could be effected by either denaturing additives or a high-pressure treatment.

Uses of Surimi in Processed Meat Products

The low-temperature setting ability of surimi has suggested its use as an effective cold binding agent for restructuring of meat, poultry, or fish pieces from trimmings (Lanier, 1985b). A freeze-dried form of surimi, trademarked Pro-bine® (Taiyo Fishery Co. Ltd., Tokyo), has been marketed in Japan specifically for this purpose. Surimi has likewise been suggested as an additive to processed red meat and poultry products to improve texture and water/fat binding because the low-temperature "setting" or surimi also leads to the development of a stronger texture and enhances water-binding properties (Lanier, 1985b; Niwa, 1992).

Yet, recent studies have determined that surimi loses its low-temperature "setting" ability when mixed with red meat at low levels (>25% beef in the mixture; Torley and Lanier, 1992). Similarly, the setting ability of surimi is disturbed when whey proteins are incorporated and could be reduced by an admixture of certain other proteins. Setting is also greatly reduced as the pH deviates from neutrality (Torley and Lanier, 1992; Nishimoto et al., 1987), further limiting the application of this property in many muscle foods because the pH of red meat and poultry typically ranges from 5.5 to 6.3 (Hultin, 1985; Daum-Thunberg et al., 1992).

Another difficulty with attempting to incorporate surimi into processed meats is the heat-stable alkaline proteinase activity which most surimi possesses (Lanier, 1986; Boye and Lanier, 1988; Lin and Lanier, 1980; Kinoshita et al., 1990). Heat-processing schedules for kamaboko and shellfish analog products commonly made with surimi are respectful of this problem, avoiding texture breakdown during cooking by maintaining process temperatures below 40°C or above 80°C, outside the active temperature range of this class of enzymes (Fig. 10–7). However, most processed meat items are cooked to an internal temperature of only about 70°C, giving proteinases which are most active at 50–70°C ample time to degrade proteins and soften texture. Although food grade inhibitors of these proteinases are available, such as beef plasma, white potato extracts, and egg white (Kinoshita et al., 1990; Hamann et al., 1990), they may not always be acceptable as ingredients for legal or marketing reasons.

A Filled-Gel Approach to the Use of Surimi in Meat/Poultry Products

Because of these difficulties in transferring the superior gelling properties of surimi to a meat/poultry system, an alternate approach to the application of surimi in processed meats can be suggested. The goal, as before, is to use the surimi to enhance the product texture and/or the water/fat-binding properties of the product. Surimi forms a gel which possesses high water and fat-binding ability (Lanier, 1986; Hamann,

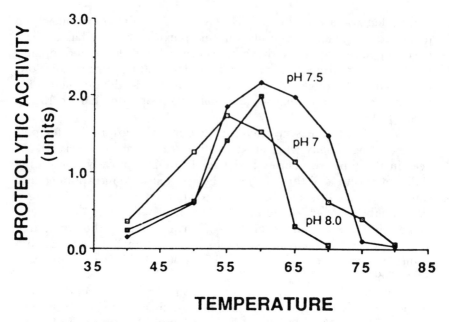

TEMPERATURE

Figure 10–7 Effects of temperature and pH on the activity of heat-activated proteinase of Atlantic menhaden.

Source: Boyle and Lanier (1988) by permission.

1987). This gel can be formed by low-temperature setting, with or without additional heat treatment, prior to being mixed with meat, incorporating the water and/or fat which would normally need to be bound within the processed meat product. This surimi gel is then added as a finely divided filler material into the raw meat batter prior to the latter being heat processed. The end result is a filled gel (gel within a gel) (Lanier, 1991) whose textural and water/fat-binding properties can be considerably enhanced by the presence of the dispersed surimi gel. The water content of the meat matrix is lowered (the normal excess has been transferred to within the surimi gel) giving rise to greater strength in this matrix. Excess fat is immobilized in the surimi matrix also (Hamann, 1987), reducing any tendency for fat to render from the product during cooking. Because the surimi gel is formed prior to incorporation in the meat batter, heating can be controlled to avoid proteinase activity and its pH remains near 7.0, ideal for gelation.

The gel produced from surimi can be very white (if manufactured from Alaska pollock or a similar lean species). When finely divided in the meat matrix, it resembles fat particles in appearance. The current consumer trend to lower-fat meat products offers a real opportunity for the surimi filler gel approach to be used for fat replacement in these products. Water becomes the primary replacement for fat, and the water is immobilized by incorporating it into the surimi gel (Lanier, 1992b).

Gellike plasteins produced from FPH might also prove useful as fillers and fat replacers, not only iln meat/poultry products but perhaps in fabricated seafood and other surimi-based products as well. Water-binding ability seems to be the key attribute for a gel to function effectively as a fat replacer in processed meats (Lanier, 1992) and it would appear from published reports that plasteins could possibly compete in this functional attribute with surimi gels (Tsai et al., 1972).

Effects of High-Pressure Treatment on Surimi Gelation

Shoji et al. (1990) reported that surimi gels can be formed at 0°C in only 10 min by application of 2.0–4.0 kbars of pressure. Evidently, transglutaminase activity is substantially enhanced by the treatment, as sodium dodecylsulfate (SDS)–urea–mercaptoethanol electrophoresis revealed substantial nondisulfide cross-linking of the myosin heavy chain in these gels. Pressure effects on surimi gelation are evident even independent of the setting phenomenon. Table 10–1 (Hayashi, 1989) compares the properties of gels prepared from Alaska pollock paste and other proteins by heat versus pressure (the latter at 25°C). Most remarkably, gels formed by pressure are much smoother and more elastic than gels formed by heat.

Low-Salt Surimi Gels

High-pressure treatment has also been shown to enhance the solubilization of myofibrillar proteins, possibly pointing to a method of preparing gelled products of lower salt content for consumers on low-sodium diets (Macfarlane and McKenzie, 1976; Suzuki and Macfarlane, 1984). Many producers of surimi-based products have already reduced the sodium content of their products by substituting potassium chloride for up to 50% of the sodium chloride required (Cooper, 1992). The addition of phosphates may also be used to reduce the total salts content needed to effect good solubilization and gelation of the surimi proteins (Okurowski, 1987).

Hultin and co-workers (Wu et al., 1991a) have recently demonstrated that the myofibrillar proteins of several fish species are quite soluble at extremely low ionic strengths, yet practically insoluble as the ionic strength is slightly raised. Because of this solubility at low ionic strength, they were able to prepare gels from red hake (*Urophycis chuss*)

Table 10–1 General characteristics of pressure-and heat-treated food proteins, including fish paste

	Treatment	
Characteristics	Pressure	Heat
Change of color	S	L
Gloss	L	S
Transparency	L	S
Density	S	L
Smoothness	L	S
Change in flavor	S	L
Hardness	S	L
Elasticity	L	L
Extensibility	L	S
Adhesiveness	L	S

Note: L = large; S = small.
Source: Hayashi (1989), by permission.

surimi in the absence of salt that were elastic enough to pass the Japanese double-fold test for surimi quality and nearly as elastic as gels prepared with salt (Table 10–2) (Wu et al., 1991b).

When a lower-sodium content is achieved by use of phosphates and substitution by potassium chloride, the low-temperature setting ability of the surimi remains undiminished (Okurowski, 1987). It is unknown whether gels prepared at extremely low ionic strengths exhibit low-temperature setting.

Lee (1991) reported that ascorbate addition can also compensate for salt reduction in maintaining the gelling properties of surimi.

Table 10–2 Protein extractable by $1M$ LiCl and properties of fish gels from various species with and without 3% added NaCl

Species	Protein Extractability (g protein/10 g dry wt.)	Fold Test Score −NaCl	Fold Test Score +NaCl	Ratio of True Strain −NaCl/+NaCl
Red hake	8.01±0.18	5	5	0.89
Cod	6.43±0.49	5	5	0.92
Dab	6.20±1.31	5	5	0.85
Pollock	5.94±1.10	3	5	0.85
Haddock	5.28±0.63	3	5	0.74
Mackerel	4.81±0.22	2	5	0.59

Source: Wu et al. (1991b).

Enhancement of Surimi Gel Strength by Ascorbic Acid and Ultraviolet Light

The main effect of presetting a gel prior to cooking is to enhance its gel strength, as the compliancy of the gel is only slightly increased (Kamath et al., 1992). As previously explained, setting results in substantial covalent cross-linking of myosin. Another method of enhancing gel strength through increased covalent cross-linking is by the addition of an oxidant, which induces the formation of intermolecular disulfide cross-linkages. Formerly, in Japan, bromates were allowed as additives for this purpose (Okada, 1961), but these have been banned due to health concerns. Bromates are commonly used as maturation agents in bread flours where they exert the same effect (Stauffer 1991). Another more acceptable oxidant also used in bread flours, ascorbic acid, has been recently explored for its applicability to surimi (Lee, 1991).

Actually, it is the oxidized form, dehydroascorbic acid (presumably formed by reaction with molecular oxygen incorporated during mixing), which is believed to be the active oxidant in these food systems (Tsen and Bushuk, 1963). Several workers have been able to demonstrate a marked increase in gel strength of cooked surimi to which ascorbic acid or ascorate was added (Lee, 1991; Yoshinaka et al., 1972; Nishimura et al., 1990). Recent work by Lee (1991) could find no effect of incorporated oxygen on the reaction, yet confirmed that disulfide bonding during heating of surimi was enhanced by ascorbate addition.

Ultraviolet radiation has also been shown to enhance the strength of heat-set surimi gels (Taguchi et al., 1989). Although ultraviolet light is known to be a pro-oxidant of many food components and can affect disulfide bonding (Cheftel et al., 1985), there is no published evidence that the mechanism of this enhancement involves sulfhydryl oxidation or disulfide interchange. Rather, ultraviolet irradiation is thought to enhance the myosin–actin association (Taguchi et al., 1989). Because ultraviolet light has poor penetration properties, the gel enhancement is a surface phenomenon and, thus, may have only limited applications.

Just as the formation of intermolecular covalent linkages among proteins during heat gelation contributes to enhanced gel strength, formation of these associations in surimi during cold storage prior to its use diminishes the gelling potential. Experimentation with several reducing agents, such as cysteine, mercaptoethanol, and sodium bisulfite has successfully demonstrated the ability to increase the number of reactive sulfhydryl groups and, thus, recover a significant portion of the loss in gelling potential caused by storage-induced denaturation and aggregation of the proteins (Jiang et al., 1986).

Supplementation of Surimi with Sarcoplasmic Proteins

Typically the water-soluble (sarcoplasmic) protein fraction of fish muscle becomes part of the waste stream of a surimi process and is discarded. Nutritionally, these proteins are equivalent to the myofibrillar proteins of muscle, yet there has been little incentive to recover them for food use. This fraction has seemed less functional than the myofibrillar and, therefore, has not been highly valued as a separate food material although it has been considered a potential source of enzymes for food and industrial processing (Haard, 1992; Stefansson and Steingrimsdottir, 1990). Earlier studies indicated that the sarcoplasmic fraction interferes with gelation of the myofibrillar proteins, thus explaining the superior gelling ability of surimi as compared to unleached fish meat (Shimizu and Nishioka, 1974; Okada, 1964). More recent studies have shown that the sarcoplasmic protein fraction might instead actually supplement the gelling properties of the myofibrillar proteins (Nishioka et al., 1990; Okazaki et al., 1986). This raised the possibility that the sarcoplasmic fraction, if separated from the nonprotein components normally leached by water in the surimi process, might be effectively added back to the surimi to enhance the overall yield of the process.

A number of processes for recovery of the soluble proteins have been examined, including electrocoagulation (Hasegawa et al., 1982), pH shifting to induce precipitation (Nishioka and Shimizu, 1983), selective binding by ion-exchange resins (Ayers and Petersen, 1985; Korhonen and Lanier, 1991), and ultrafiltration (French and Pedersen, 1990). Recovery by the latter two methods yields a relatively undenatured protein which has been shown to effectively replace 10% or more of the protein in surimi with no adverse effects to flavor, color, gel-forming ability, or stability (Korhonen and Lanier, 1992). Of these two, ultrafiltration is presently the more commercially proven and cost-effective recovery technique.

STABILIZATION OF FISH PROTEINS TO FROZEN STORAGE AND DRYING

For ingredient materials to be handled easily by food manufacturers, they must be available in a form which can be stored in a stable state for reasonable periods of time prior to use. The low heat stability of fish proteins does not lend them to stabilization by current aseptic packaging methods involving heat sterilization, but irradiation may prove to be a viable and acceptable cold-sterilization process in the future (Gorga and

Ronsivalli, 1988). Meanwhile, fish proteins will likely be handled and stored in the frozen or dried form.

Stabilization of fish proteins in frozen storage requires the removal or inactivation of denaturants and/or the protection of the proteins from remaining denaturant actions. The water-leaching step of surimi manufacture accomplishes the first of these objectives to some degree by decreasing the lipid content and removing deleterious enzyme systems and/or their cofactors (Holmquist, 1982). The addition of cryoprotectant additives accomplishes the second objective. Typically, low-molecular-weight sugars and sugar alcohols have been used as protein-protective agents, whereas phosphates were added for pH adjustment (the proteins are more stable in the neutral range) and as antioxidants. The range of cryoprotectant materials available is increasing as our knowledge of cryoprotectant mechanisms has increased. Low-molecular-weight compounds seem to exert their protective effects by a solute exclusion principle. Conversely, oligosaccharides such as maltodextrins and Polydextrose® may act primarily to raise the glass transition temperature, ensuring that the hydrated system exists as an immobilized, stable glass at conventional cold-storage temperatures (MacDonald and Lanier, 1991). These compounds exhibit little sweetness, a possible advantage in some applications over sugars and polyols. Several of the starch hydrolysates may, however, interfere with the gelling properties of fish proteins such that applications in surimi (but not most mince applications) could be limited (Park et al., 1988; Wu et al., 1985).

Sugars have also proven effective as "dryoprotectants" in the dehydration of surimi by spray- or freeze-drying to form a functional protein powder (Niki et al., 1992). In this case, the mechanism for protection seems opposite to that for cryoprotection: Rather than being preferentially excluded from the protein surface, the protective compound replaces water molecules at the surface of the proteins, thus preventing dehydration-induced conformational changes and protein–protein interactions that result in the loss of gelling ability (MacDonald and Lanier, 1991; Crowe et al., 1990). Because dried materials are more easily stored and handled, the future competitiveness of surimi as a food ingredient may well depend on the development of cost-efficient drying techniques which do not detract from its stability and gelling functionality.

FISH GELATINS: A CASE EXAMPLE IN MARKETING FISH PROTEINS

Fish gelatins are an excellent example of how niche marketing of a fish protein ingredient can succeed, given that it possesses unique functional properties and is properly undergirded by technical support to users. Fish gelatins derived from the collagenous skins of cold-water species are less prone to gel than mammalian gelatins but are said to be excellent emulsion stabilizers, crystal growth inhibitors, and film-forming agents (Norland, 1990). Moreover, they can replace mammalian gelatins in applications where religious prohibitions to these exist. Presently, there are investigations into extending the range of gelling properties from fish gelatins by utilizing warmer-water fishes as raw material (Anonymous, 1991b).

Manufacturers of fish gelatins typically issue technical specification sheets for each type of product which they offer, itemizing the functional properties, composition, and microbiological characteristics. Technically trained staff are available to answer questions by phone and to assist in testing of the product in actual plant applications. Sales personnel are also trained to ferret out new applications for fish gelatins not currently being met by the competitive mammalian products.

This operational model is not unique to fish gelatins but is common to the marketing of all specialty food ingredients. The fundamental basis of the marketing approach is an emphasis on the functionality of the material. This functionality must be repeatably measurable in controlled tests with accepted methods and equipment to provide a means of quality control and product specifications. These model tests results must then be able to be correlated with the benefits which can be realized in each application of the ingredient: Improved cost, yield, texture, flavor, appearance, or machinability are marketable benefits to be explored. Finally, technical staff must be retained by the company to assist customers in using the ingredient and exploring potential new applications.

This approach to the marketing of other fish proteins has been sadly lacking in the past and is the likely cause of the many failures experienced. The success of surimi on the world protein market has heretofore derived from its value as an exclusive food base for certain specialty foods, notably kamaboko products and shellfish analogs. Price increases have driven many manufacturers of these products to seek alternatives to surimi, and many nonfish protein suppliers are rising to the challenge (Anonymous, 1990, 1991b). Surimi and other fish protein manufacturers must carefully consider whether they wish to continue producing for a strictly commodity market, subject to the attacks of alternative sources and suppliers. An attractive alternative might be to market fish proteins as specialty food ingredients. This will, however, require a commitment to gaining technical understanding of their products and processes so that they may identify and meet the specialized needs of different food manufacturers.

One technical development which should be helpful in this regard is the recent development in the United States of standardized testing methods for surimi quality (Lanier et al. 1991). This will provide suppliers a means of accurately assessing the functional properties of fish proteins destined for gelling applications. It also yields data suitable for computerized least-cost formulation (by linear programming techniques) of products dependent on the gelling properties of fish proteins (Lanier, 1992a). This technique can be used to excellent advantage with any type of food to determine or demonstrate the economic and functional viability of particular ingredients vis-à-vis competitive ingredients.

EXTENDING THE RANGE OF FUNCTIONAL INGREDIENTS FROM FISH

The question may logically be asked: How might processors develop a product range of fish protein ingredients that display a variety of functional properties? The fish gelatin example is again a good model. Fish gelatin itself is differentiated from its main competition, mammalian gelatins, by its lower gelling properties, which are desired in particular markets. As explained earlier, this property is a function of its differing amino acid content, which is a function of species and habitat temperature. Thus, species inhabiting warmer waters yield gelatins more similar in gelling properties to mammalian gelatins (Norland 1990). Process variations may also be employed to manipulate the gelling and solubility properties further, by adjusting the degree of collagen hydrolysis, distribution of molecular weights, and so on. The gelatin may be sold in liquid or dried form, and the dried form may vary in mesh size. These variations enable the manufacturers to tailor a product to each specific application.

Similarly, the stability and gelling properties of fish myofibrillar proteins are unique as compared to mammalian myofibrillar proteins largely due to fish being cold blooded. Myosin isoforms exist between species and can shift within a species due to environmental and perhaps other factors, such as age, habitat, feed, and so on (Hwang et al., 1991; Johnston et al., 1990). Contents of dark and light muscle also vary among species (particularly comparing pelagics to demersal species), affecting not only gelling properties but also color and flavor characteristics (Shimizu et al., 1992; Hultin, 1985). Given the large number of potential species and habitats, these two factors alone enable substantial variation in functional properties of surimi or similar protein ingredient products which can be made from these fishes. To this may be added variations in processing techniques which may affect the fractional contents of myofibrillar, sarcoplasmic, and connective tissue contents, and even relative weighting of the subfractions of each. Product forms, fresh, frozen, and dried, with countless variations introduced by way of additives, particle size, and freezing technique introduce additional product options.

SUMMARY

Many agricultural commodities are presently fractionated into various components which serve as ingredients for a host of formulated foods. Fish can be a source of highly functional food protein which, if properly marketed with adequate technical support, can meet specific ingredient needs within the food industry. These needs should be specifically identified so they may become the market "pull" for further technical developments in the manufacture of protein food ingredients from fish.

Figure 10–8 summarizes the processing and fractional refining of fish into functional protein food ingredients. Recognize that this should merely be a starting point for exploration of fish as a protein ingredient source. Commodities which have been successfully exploited as sources of functional food proteins continue to be examined in light of new market needs. For example, specialty dairy proteins have been developed for every conceivable product requirement, and such development and subsequent technical marketing has propelled countries such as Denmark and New Zealand to the forefront as food ingredient suppliers, in addition to their established reputations as sources of milk, cheese, and butter. Yet, new dairy protein ingredients are even yet being developed, such as high α-lactalbumin or β-lactoglobulin whey protein isolates processed by novel ion-exchange/immobilized affinity column methods (Anonymous 1991a; Jang and Swaisgood, 1990).

The challenge is for the fishing industry to think and market functional proteins on a firm technical basis. Cross-fertilization from research on competing proteins should be encouraged, as advances in the science of utilizing of one commodity can often be transferred to another. In this regard, current research into the creation of high-functionality dairy proteins, by selective enzymatic cleavage and fractionation of functional protein domains, may be a model for future developments in the functionality of fish proteins (Chen and Swaisgood, 1991). Genetic manipulation is on the horizon also; for example, the "antifreeze protein" which inhibits ice crystal formation and is, thus, responsible for the cold endurance of winter flounder has been successfully cloned into other species (Fletcher et al., 1988).

Meanwhile, a sufficient strong research base already exists for diver-

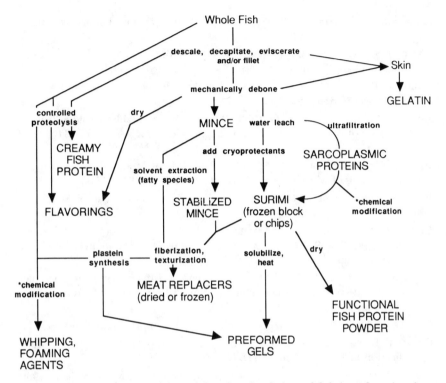

Figure 10–8 The processing and fractional refining of fish into functional protein food ingredients.

sification of product functionality. It remains for the fish industry to capitalize on it.

REFERENCES

ANONYMOUS. 1988. "A Trial to Establish Technical and Commercial Viability of Cod Frame Mince Recovery. Canadian Industry Report of Fisheries and Aquatic Science, No. 197.

ANONYMOUS. 1990. "Surimi Based Foods Using Alaco Surimi Plus (Whey Protein Concentrate)." Technical application A604, Milk Products (N.Z.) Limited, UK.

ANONYMOUS, 1991a. Albuvir (whey protein isolate). Technical brochure, Union Laitiere Normande, France.

ANONYMOUS. 1991b. *Fish Gelatin*. Technical brochure. Croda Colloids Ltd., England.

ARAI, S., and FUJIMAKI, M. 1978. "The Plastein Reaction. Theoretical Basis." *Ann. Nutr. Aliment.* 32:701–707.

AYERS, J. S., and PETERSON, M. J. 1985. "Whey Protein Recovery Using a Range of Novel Ion-Exchangers." *New Zealand J. Dairy Sci. Technol.* 20:129–142.

BALDWIN, R. E., and SINTHAVALAI, S. 1974. "Fish Protein Concentrate Foam." *J. Food Sci.* 39:880–882.

BOYE, S. W., and LANIER, T. C. 1988. "Effects of Heat-Stable Alkaline Protease Activity of Atlantic Menhaden on Surimi Gels." *J. Food Sci.* 53:1340–1342, 1398.

BREKKE, C. J., and EISELE, T. A. 1981. "The Role of Modified Proteins in the Processing of Muscle Foods." *Food Technol.* 35(5):231–234.

BROOKS, GLADDEN M. 1989. Personal communication, Mississippi State University.

CHEFTEL, J. C., CUQ, J-L., and LORIENT, D. 1985. "Amino Acids, Peptides, and Proteins." P. 339. In *Food Chemistry*, edited by O. W. Fennema, New York: Marcel Dekker.

CHEN, S. X., and SWAISGOOD, H. E. 1991. "Characteristics of a Structural Domain Prepared from β-Lactoglobulin by Limited Proteolysis with Immobilized Trypsin. *J. Dairy Sci.* 74 (Suppl. 1):102.

COOPER, G. M. 1992. Personal communication. Morton Salt Co., Woodstock, IL.

CREAMER, L. K., JIMENEZ-FLORES, R., and RICHARDSON, T. 1988. "Genetic Modification of Food Proteins." *TIBTECH* 6:163–169.

CROWE, J. H., CARPENTER, J. F., CROWE, L. M., and ANCHORDOGUY, T. J. 1990. "Are Freezing and Dehydration Similar Stress Vectors? A Comparison of Modes of Interaction of Stabilizing Solutes with Biomolecules." *Cryobiology* 27:219.

DAUM-THUNBERG, D. L., FOEGEDING, E. A., and BALL, H. R., JR. 1992. "Rheological and Water-Holding Properties of Comminuted Turkey Breast and Thigh: Effects of Initial pH." *J. Food Sci.* 57:333–338.

FLETCHER, G. L., SHEARS, M. A., KING, M. J., DAVIES, P. L., and CHEW, C. L. 1988. "Evidence for Antifreeze Protein Gene Transfer in Atlantic Salmon (*Salmo salar*)." *Can. J. Fish Aquat. Sci.* 45:352.

FRENCH, J. S., and PEDERSEN, L. D. 1990. "Properties and Stability of Surimi Combined with Process-Water Protein Concentrates." Pp. 95–102. In *Chilling and Freezing of New Fish Products*, Intl. Inst. Refrig., Paris.

GOLDHOR, S. H., and REGENSTEIN, J. M. 1988. "U.S. Fishery Byproducts: A Selective Update and Review. Pp. 213–221. In Proc. 12th Ann. Trop. Subtrop. Fish. Technol. Soc. Amer., Orlando, FL: Florida Sea Grant Prog. SGR–92, Gainesville, FL.

GORGA, C., and ROSIVALLI, L. J. 1988. *Quality Assurance of Seafood.* New York: Van Nostrand Reinhold.

GREENBERG, C. S., BIRCHBICHLER, P. J., and RICE, R. H. 1991. "Transglutaminases: Multifunctional Cross-Linking Enzymes That Stabilize Tissues." *FA-SEB J.* 5:3071–3077.

GWINN, S. E. 1992. "Development of Surimi Technology in the United States." Pp. 23–40. In *Surimi Technology.* edited by T. C. Lanier and C. M. Lee. New York: Marcel Dekker.

HAARD, N. F. 1992. "A Review of Proteolytic Enzymes from Marine Organisms and Their Application in the Food Industry." *J. Aquatic Food Prod. Technol.* 1:17–35.

HALLERMAN, E. M., KAPUSCINSKI, A. R., HACHETT, P. B., JR., FARRAS, A. J., and GUISE, K. S. 1990. "Gene Transfer in Fish." In Pp. 35–49. *Advances in Fisheries Technology and Biotechnology for Increased Profitability,* Lancaster, PA: Technomic Publishing Co.

HAMANN, D. D. 1987. "Instrumental Texture Measurements for Processed Meat Products." *Proc. Ann. Recipr. Meat Conf. Nat. Live Stock Meat Bd.* 40:19–23.

HAMANN, D. D., AMATO, P. M., WU, M. C., and FOEGEDING, E. A. 1990. "Inhibition of Modori (Gel Weakening) in Surimi by Plasma Hydrolysate and Egg White." *J. Food Sci.* 55:665–669, 794.

HASEGAWA, H., WATANABE, H., and TAKAI, R. 1982. "Methods of Recovery of Fish Muscle Water-Soluble Protein by Electrocoagulation." *Nippon Suisan Gakkaishi.* 48:65–68.

HASHIMOTO, A., and ARAI, K. 1984. "Temperature Dependence of Mg-ATPase Activity and its Ca-Sensitivity of Fish Myofibrils." *Nippon Suisan Gakkaishi.* 50:853–863.

HASHIMOTO, A., KOBAYASHI, A., and ARAI, K. 1982. Thermostability of fish myofibrillar Ca-ATPase and adaption to environmental temperature. Nippon Suisan Gakkaishi 48(5):671–684.

HASTINGS, R. J. 1989. Comparison of the Properties of Gels Derived from Cod Surimi and from Unwashed and Once-Washed Cod Mince." *Internat. J. Food Sci. Technol.* 24:93–102.

HAYASHI, R. 1989. "Application of High Pressure to Food Processing and Preservation: Philosophy and Development. Pp. 815–826. In *Engineering and Food,* edited by W. E. L. Spiess and H. Schubert. Vol. 2. London: Elsevier Applied Science.

HOLDEN, C. 1971. "Fish Flour: Protein Concentrate Has Yet to Fulfill Expectations." *Science* 173:410.

HOLMES, K. (ed). 1987. "Surimi: It's American Now." Anchorage, AK: Alaska Fisheries Development Foundation.

HOLMQUIST, J. F. 1982. "*Interrelations Between Salt-Extractable Protein, Actomyosin, Calcium ATPase Activity, and Kamaboko Quality Prepared from Frozen Red Hake Fillets, Mince, and Surimi.* M. S. thesis, University of Massachusetts.

HORISBERGER, M. 1979. "Lessons from the Past for Better Future Utilization of Fish Resources." Pp. 41–49. *Nestle Research News.* Lausanne, Nestle Products Technical Assist. Co. Ltd.

HULTIN, H. O. 1985. "Characteristics of Muscle Tissue." P. 753. In *Food Chemistry*, edited by O. R. Fennema, New York: Marcel Dekker.

HULTIN, H. O. 1987. "Factors Responsible for Lipid Oxidation in Fish Muscle." Pp. 185–224. In *Proc. Conf. Fatty Fish Utilization: Upgrading from Feed to Food*. University of North Carolina, Raleigh, NC, SeaGrant Publ. 88–04.

HWANG, G.-C., OCHIAI, Y., WATABE, S., and HASHIMOTO, K. 1991. "Changes of Carp Myosin Subfragment-1 Induced by Temperature Acclimation." *J. Com. Physiol. B.* 161:141–146.

IN, T. 1990. "Seafood Flavourants Produced by Enzymatic Hydrolysis." Pp. 425–436. In *Advances in Fisheries Technology and Biotechnology for Increased Profitability*. Lancaster, PA: Technomic Publishing.

JANG, H. D., and SWAISGOOD, H. E. 1990. "Analysis of Ligand Binding and β-Lactoglobulin Denaturation by Chromatography on Immobilized Trans-Retinal." *J. Dairy Sci.* 73:2067–2074.

JIANG, S.-T., LAN, C. C., and TSAO, C.-Y. 1986. "New Approach to Improve the Quality of Minced Fish Products from Freeze-Thawed Cod and Mackerel." *J. Food Sci.* 51:310–312, 351.

JOHNSTON, I. A., and GOLDSPINK, G. 1975. "Thermodynamic Activation Parameters of fish Myofibrillar ATPase Enzyme and Evolutionary Adaptations to Temperature." *Nature* 257:620.

JOHNSTON, I. A., FLEMING, J. D., and CROCKFORD, T. 1990. "Thermal Acclimation and Muscle Contractile Properties in Cyprinid Fish." *Amer. J. Physiol: Regulatory, Integr. Comp. Physiol.* 28: R231–R236.

KAMATH, G. G., LANIER, T. C., FOEGEDING, E. A., and HAMANN, D. D. 1992. "Non-disulfide Covalent Cross-Linking of Myosin Heavy Chain in 'Setting' of Alaska Pollock and Atlantic Croaker Surimi." *J. Food Biochem.* 16:151–172.

KEAY, J. N. (ed.). 1976. *Proceedings of the Conference on the Production and Utilization of Mechanically Recovered Fish Flesh*. 108 Pages. Aberdeen, Scotland: Ministry of Agriculture, Fisheries and Food.

KELLEHER, S. D., SILVA, L. A., HULTIN, H. O., and WILHELM, K. A. 1992. "Inhibition of Lipid Oxidation During Processing of Washed Minced Atlantic Mackerel." *J. Food Sci.* 57:1103–1108, 1119.

KIMURA. I., SUGIMOTO, M., TOYODA, K., SEKI, N., ARAI, K., and FUJITA, T. 1991. "A Study on the Cross-Linking Reaction of Myosin in Kamaboko 'suwari' Gels." *Nippon Suisan Gakkaishi.* 57:1389–1396.

KINOSHITA, M., TOYOHARA, H., and SHIMIZU, Y. 1990. "Proteolytic Degradation of Fish Gel (Modori Phenomenon) During Heating Process." In *Chilling and Freezing of New Fish Products*. Paris: Intl. Inst. Refrig.

KISHI, H., NOZAWA, H., and SEKI, N. 1991. "Reactivity of Muscle Transglutaminase on Carp Myofibrils and Myosin B." *Nippon Suisan Gakkaishi.* 57:1203–1210.

KNIGHT, M. K. 1990. "New Meat Ingredients." Pp. 129–134. In *Food Ingredients Exhibition Britain—Seminar Proceedings*. London: Eppoconsult Publishers.

KOLBE, E. R., HSU, C. K., LANIER, T. C., MACDONALD, G. A., MORRISSEY, M. T., and SIMPSON, R. 1992. "Product Alternatives for Pacific Whiting." *Proceedings of a Pacific Whiting Workshop on Harvesting, Processing, Marketing and Quality Assurance Programs*. Astoria: Oregon State University.

KORHONEN, R. W., and LANIER, T. C. 1991. "Recovery of Surimi Leachwater Proteins." *Proc. Third Seafood Technol. and Regulatory Developments Conference: Seafood Waste Issues in the 1990's*. Washington, D.C.: New Orleans National Fisheries Institute.

KORHONEN, R. W., and LANIER, T. C. 1992. Unpublished data. North Carolina State University, Raleigh, NC.

KURTH, L., and ROGERS, P. J. 1984. "Transglutaminase Catalyzed Cross-Linking of Myosin to Soya Protein, Casein and Gluten." *J. Food Sci.* 49:573.

LANIER, T. C. 1985a. "Menhaden: Soybean of the Sea. University of North Carolina, Raleigh, NC, Publ. No. UNC-SG–85–02.

LANIER, T. C. 1985b. "Fish Protein in Processed Meats." *Proc. Ann. Recipr. Meat Conf., Nat. Live Stock Meat Bd.* 38:129–134.

LANIER, T. C. 1986. "Functional Properties of Surimi." *Food Technol.* 40(3):107–114, 124.

LANIER, T. C. 1991. "Interactions of Muscle and Nonmuscle Proteins Affecting Heat-Set Gel Rheology." In Pp. 268–284. *Interactions of Food Proteins*, edited by N. Parris and R. Barford. Washington, DC: American Chemical Society. Symposium Series 454.

LANIER, T. C. 1992a. "Measurement of Surimi Composition and Functional Properties. Pp. 123–163. In *Surimi Technology*. edited by T. Lanier and C. Lee. New York: Marcel Dekker.

LANIER, T. C. 1992b. Unpublished data. North Carolina State University, Raleigh.

LANIER, T. C., HART, K., and MARTIN, R. E. 1991. In *A Manual of Standard Methods for Measuring and Specifying the Properties of Surimi*. UNC Sea Grant Publication SG-91-01. Washington, DC: National Fisheries Institute.

LANIER, T. C., MANNING, P. K., ZETTERLING, T., and MACDONALD, G. A. 1992. "Process Innovations in Surimi Manufacture." Pp. 167–180. In *Surimi Technology*, edited by T. C. Lanier and C. M. Lee. New York: Marcel Dekker.

LEE, H. G. 1991. *Mechanism of Gel-Strengthening Effect of Sodium Ascorbate in Surimi Gel*. M. S. thesis, University of Rhode Island, Kingston, RI.

LIN, T. S., and LANIER, T. C. 1980. "Properties of an Alkaline Protease from the Skeletal Muscle of Atlantic Croaker." *J. Food Biochem.* 4:17–28.

MACDONALD, G. A., and LANIER, T. C. 1991. "Carbohydrates as Cryoprotectants for Meats and Surimi." *Food Technol.* 45(3):150–159.

MACDONALD, G. A., LELIEVRE, J., and WILSON, N. D. C. 1990a. "Strength of

Gels Prepared from Washed and Unwashed Minces of Hoki (*Macruronus novaezelandiae*) Stored in Ice." *J. Food Sci.* 55:976–978, 982.

MacDONALD, G. A., WILSON, N. D. C., and LANIER, T. C. 1990b. "Stabilised Mince: An Alternative to the Traditional Surimi Process." Pp. 69–76. In *Chilling and Freezing of New Fish Products*, Intl. Inst. Refrig., Paris.

MACFARLANE, J. J., and McKENZIE, I. J. 1976. "Pressure-Induced Solubilization of Myofibrillar Proteins." *J. Food Sci.* 41:1442–1446.

MACKIE, I. M. 1983. "New Approaches in the Use of Fish Proteins." Pp. 215–262. In *Developments in Food Proteins—2*, edited by B.J.F. Hudson. New York: Applied Science Publishers.

MARTIN, R. E. (ed.). 1972. In *Proceedings of the First Technical Seminar on Mechanical Recovery and Utilization of Fish Flesh*. Washington, DC: National Fisheries Institute and U. S. Department of Commerce, National Marine Fisheries Service.

MARTIN, R. E. (ed.). 1974. In *Proceedings of the Second Technical Seminar on Mechanical Recovery and Utilization of Fish Flesh*. Washington, DC: National Fisheries Institute and U. S. Department of Commerce, National Marine Fisheries Service.

MARTIN, R. E. (ed.). 1980. In *Proceedings of the Third National Technical Seminar on Mechanical Recovery and Utilization of Fish Flesh*. Washington, D.C.: National Fisheries Institute and U. S. Department of Commerce, National Marine Fisheries Service.

MAURON, J. 1979. "The Analytical, Nutritional and Toxicological Implications of Protein Food Processing." Pp. 51–58. *Nestle Research News*, Lausanne: Nestle Products Technical Assist. Co. Ltd.

MONTEJANO, J. G., HAMANN, D. C., and T. C. LANIER, 1984. "Thermally Induced Gelation of Selected Comminuted Muscle Systems—Rheological Changes During Processing, Final Strengths and Microstructure." *J. Food Sci.* 49:1496–1505.

NIKI, H., MATSUDA, Y., and SUZUKI, T. 1992. "Dried Forms of Surimi." Pp. 209–243. In *Surimi Technology*, edited by T. Lanier and C. Lee. New York: Marcel Dekker.

NISHIMOTO, S. I., HASHIMOTO, A., SEKI, N., KIMURA, I., TOYODA, K., and FUJITA, T. 1987. "Influencing Factors on Changes in Myosin Heavy Chain and Jelly Strength of Salted Meat Paste from Alaska Pollock During Setting." *Nippon Suisan Gakkaishi*. 53:2011–2020.

NISHIMURA, K., OHTSURU, M., and NIGOTA, K. 1990. "Mechanism of Improvement Effect of Ascorbic Acid on the Thermal Gelation of Fish Meat." *Nippon Suisan Gakkaishi*. 56:959.

NISHIOKA, F., and SHIMIZU, Y. 1983. "Recovery of Proteins from Washings of Minced Fish by pH Shifting Method." *Nippon Suisan Gakkaishi*. 49:795–800.

NISHIONKA, F., TOKUNAGA, T., FUJIWARA, T., and YOSHIOKA, S. 1990. "Development of New Leaching Technology and a System to Manufacture High Quality Frozen Surimi." Pp. 123–32. In *Chilling and Freezing of New Fish Products,* Intl. Inst. Refrig., Paris.

NIWA, E. 1992. "Chemistry of Surimi Gelation." Pp. 389–428. In *Surimi Technology,* edited by T. Lanier and C. Lee. New York: Marcel Dekker.

NORLAND, R. E. 1990. "Fish Gelatin." Pp. 325–333. In *Advances in Fisheries Technology and Biotechnology for Increased Profitability.* Edited by M. N. Voigt and J. R. Botta. Lancaster, PA: Technomic Publishing Co.

OKADA, M. 1961. "The Effect of Oxidants on Jelly Strength of 'Kamaboko.' " *Nippon Suisan Gakkaishi.* 27:203–208.

OKADA, M. 1964. "Effect of Washing on the Jelly Forming Ability of Fish Meat." *Nippon Suisan Gakkaishi.* 30:255–261.

OKAZAKI, E., KANNA, K., and SUZUKI, T. 1986. "Effect of Sarcoplasmic Protein on Rheological Properties of Fish Meat Gel Formed by Retort Heating." *Nippon Suisan Gakkaishi.* 52:1821–1827.

OKLAND, O. 1986. "High Quality Raw Material: A Condition for Success." *Fiskets Gang (Norway).* 9(17):267–268.

OKUROWSKI, V. 1987. *Effects of Phosphate Addition on the Heat Gelation Properties of Surimi.* M. S. thesis, North Carolina State University, Raleigh, NC.

PARISER, E. R., WALLERSTEIN, M. B., CORKERY, C. J., and BROWN, N. L. 1978. In *Fish Protein Concentrate: Panacea for Protein Malnutrition?"* Cambridge, MA: MIT Press.

PARK, J. W., LANIER, T. C., and GREEN, D. P. 1988. "Cryoprotective Effects of Sugar, Polyols, and/or Phosphates on Alaska Pollock Surimi." *J. Food Sci.* 53:1–3.

POMERANZ, Y. 1992. "Modified Muscle Proteins." Pp. 217–221. In *Functional Properties of Food Components,* San Francisco, CA: Academic Press.

SAEKI, H., WAKAMEDA, A., ICHIHARA, Y., and SASAMOTO, Y. 1989. "Effect of $CaCl_2$ on the Elution of Cross-Linking Factor in Alaska Pollock." *Nippon Suisan Gakkaishi.* 55:1867.

SANO, T., NOGUCHI, S. F., TSUCHIYA, T., and MATSUMOTO, J. J. 1986. "Contribution of Paramyosin to Marine Meat Gel Characteristics." *J. Food Sci.* 51:946.

SANO, T., NOGUCHI, S. F., TSUCHIYA, T., and MATSUMOTO, J. J. 1989. "Paramyosin–Myosin–Actin Ineractions in Gel Formation of Invertebrate Muscle. *J. Food Sci.* 54:796–799, 842.

SEAL, R. 1977. "Soya Products: A Food Processor's Guide." *Chem. Ind. (London)* 11:441–446.

SHENOUDA, S. Y. 1980. "Theories of Protein Denaturation During Frozen Storage of Fish Flesh." Edited by C. O. Chichester, E. M. Mrak and G. F. Stewart. *Adv. Food Res.* 26:275–311. New York: Academic Press.

SHIMIZU, Y., and NISHIOKA, F. 1974. "Species Variations in Heat Coagulation Properties of Fish Actomyosin–Sarcoplasmic Protein Systems." *Nippon Suisan Gakkaishi.* 40:267–270.

SHIMIZU, Y., TOYOHARA, H., and LANIER, T. C. 1992. "Surimi Production from Fatty and Dark-Fleshed Fish Species. Pp. 181–208. In *Surimi Technology*, edited by T. Lanier and C. Lee. New York: Marcel Dekker.

SHOJI, Y. 1990. "Creamy Fish Protein. Pp. 87–93. In *Proc. Internat. Conf. on Fish By-Products*, Anchorage, AK: Alaska Sea Grant College Prog. Report No. 90–07.

SHOJI, T., SAEKI, H., WAKAMEDA, A., NAKAMURA, M. and NONAKA, M. 1990. Gelation of salted paste of Alaska pollock by high hydrostatic pressure and change in myofibrillar protein in it. *Nippon Suisan Gakkaishi* 56: 2069–2076.

SPINELLI, J., KOURY, B., GRONINGER, H., JR., and MILLER, R. 1977. "Expanded Uses for Fish Protein from Underutilized Species." *Food Technol.* 31(5): 184–187.

STAUFFER, C. E. 1991. "Oxidants." Pp. 1–40. In *Functional Additives for Bakery Foods*. New York: Van Nostrand Reinhold.

STEFANSSON, G., and STEINGRIMSDOTTIR, U. 1990." Application of Enzymes for Fish Processing in Iceland." Pp. 237–250. In *Advances in Fisheries Technology and Biotechnology for Increased Profitability*. Edited by M. N. Voigt and J. R. Botta. Lancaster, PA: Technomic Publishing Co.

SUZUKI, T. 1981. "Frozen Minced Meat (Surimi). P. 121. In *Fish and Krill Protein: Processing Technology*. London: Applied Science Publisher.

SUZUKI, T., and MACFARLANE, J. J. 1984. "Modification of the Heat-Setting Characteristics of Myosin by Pressure Treatment." *Meat Sci.* 11:263–274.

SWAFFORD, T. C., BABBITT, J., REPPOND, K., HARDY, A., RILEY, C. C., and ZETTERLING, T. K. A. 1985. "Surimi Process Yield Improvements and Quality Contribution by Centrifuging." *Proceedings of the International Symposium on Engineered Seafood Including Surimi*, edited by R. E. Martin. Washington, DC: National Fisheries Institute.

TAGUCHI, T., ISHIZAKI, S., TANAKA, M., NAGASHIMA, Y., and AMANO, K. 1989. "Effect of Ultraviolet Irradiation on Thermal Gelation of Muscle Pastes." *J. Food Sci.* 54: 1438–1440, 1465.

TANAKA, M., SUZUKI, K., and TAGUCHI, T. 1983. "Recovery of Proteins as a Spun Product from Sardine Viscera and Heads." *Nippon Suisan Gakkaishi.* 49:1701–1705.

TANNENBAUM, S. R., AHERN, M., and BATES, R. P. 1970a. "Solubilization of Fish Protein Concentrate. 1. An Alkaline Process." *Food Technol.* 24:604–607.

TANNENBAUM, S. R., BATES, R. P., and BRODFELD, L. 1970b. "Solubilization of Fish Protein Concentrate. 2. Utilization of the Alklaine-Process Product." *Food Technol.* 24:607–609.

TORLEY, P. J., and LANIER, T. C. 1992. "Setting Ability of Salted Beef–Pollock Surimi Mixtures." Pp. 305–316. In *Seafood Science and Technology*, edited by E. G. Bligh. Oxford: Fishing News Books.

TSAI, S-J., YAMASHITA, M., ARAI, S., and FUJIMAKI, M. 1972. "Effect of Substrate

Concentration on Plastein Productivity and Some Rheological Properties of the Products." *Agric. Biol. Chem.* 36:1045–1049.

TSEN, C. C., and BUSHUK, W. 1963. "Changes in Sulfhydryl and Disulfide Contents of Doughs During Mixing Under Various Conditions." *Cereal Chem.* 40:339.

UTSUMI, S., and KITO, M. 1991. "Improvement of Food Protein Functions by Chemical, Physical, and Biological Modifications." *Comments Agric. Food Chem.* 2:261–278.

WU, M. C., LANIER, T. C., and HAMANN, D. D. 1985. "Thermal Transitions of Admixed Starch/Fish Protein Systems During Heating." *J. Food Sci.* 50:20–25.

WU, Y.-J., ATALLAH, M. T., and HULTIN, H. O. 1991a. "The Proteins of Washed, Minced Fish Muscle Have Significant Solubility in Water." *J. Food Biochem.* 15:209–218.

WU, Y.-J., ATALLAH, MT., HULTIN, H. O., and BAKIR, H. 1991b. "Relation of Water Solubility of Fish Muscle Proteins to Gel Formation in the Absence of Salt." Pp. 275–284. In *Proc. Tropical Subtropical Fish. Technol. Conf.*, Raleigh, NC: Florida Sea Grant Prog. SGR–110, Gainesville, FL.

YOSHINAKA, R., SHIRAISHI, M., and IKEDA, S. 1972. "Effect of Ascorbic Acid on the Gel Formation of Fish Meat." *Nippon Suisan Gakkaishi.* 38:511–515.

11

Seafood Protein in Human and Animal Nutrition

Shi-Yen Shiau

NUTRITIONAL FUNCTIONS AND REQUIREMENTS OF PROTEIN

In the body, protein is used for the structural formation of cells and tissues, the production of various essential compounds such as enzymes, antibodies, hormones, and protein mediators for regulating fluid and electrolyte balances and well as blood neutrality. It can also be utilized as an energy source (1 g of protein provides 4 kcal).

Dietary protein is quantitatively used for depositing new protein in tissues of pregnant women, infants, and children, for the protein secretion in milk of lactating women, and for the maintenance of body protein synthesis in adults. Thus, inadequate protein intake causes diminished protein content in cells and organs and deterioration in the cellular capacity to perform normal functions. This leads to increased morbidity and mortality. On the other hand, excess protein intake of physiologic need is also disadvantageous (Young and Pellet, 1987). Therefore, an adequate diet must contain an appropriate level of protein for the assurance of long-term health.

It is generally accepted that the relative concentration of essential amino acids is the major factor determining the nutritional value of food protein. Most animal proteins have a satisfactory essential amino acid pattern in relation to the amino acid requirements. Therefore, animal proteins are of high quality. On the other hand, vegetable protein may be of lower nutritional value because they tend to be limiting in one or more of the essential amino acids.

SEAFOOD VERSUS FARM ANIMALS AS THE SOURCE OF ANIMAL PROTEIN IN HUMAN DIETS

Aquatic species convert practical feeds into body tissue more efficiently than do farm animals. Cultured catfish gain approximately 0.84 g of weight per gram of practical diet, whereas chickens, the most efficient warm-blooded food animal, gain about 0.48 g of weight per gram of diet (Table 11–1) (Lovell 1988). The reason for the superior food conversion efficiency of aquatic species is that they are able to assimilate diets with higher percentages of protein, apparently because of their lower dietary energy requirement. Fish, however, do not hold an advantage over chickens in protein conversion; as shown in Table 11–1, poultry convert dietary protein to body protein at nearly the same rate as fish. The primary advantage of fish over land animals is lower energy cost of protein gain rather than superior food conversion efficiency. Protein gain per megacalorie of energy consumed is 47 g for channel catfish versus 23 g for the broiler chicken.

Unfortunately, a total energy (physiological and fossil) budget for the production of protein from freshwater fish culture systems has not been developed as precisely as budgets for terrestrial animal and plant proteins. The fossil energy required to grow channel catfish in ponds has been estimated to be similar to that needed for broiler chicken production (Lovell et al., 1978); for example, chickens require heating and ventilation and catfish require pumped water and aeration. Processing methods for channel catfish and broiler chickens are also similar. They include transport from the production site to a nearby processing site, slaughter, and ice-packing or freezing the dressed carcass. Assuming that the fossil energy requirements for producing and processing catfish and chickens are similar, the lower metabolic energy requirement for protein synthesis by catfish (Table 11–1) makes them a more energy-efficient source of protein. Other land animals require more fossil and dietary energy to produce body protein than chickens.

The percentage of edible lean tissue in aquatic species is appreciably greater than that in beef, pork, or poultry (Table 11–2). For example,

Table 11-1 Comparison of efficiency of utilization of feed dietary protein and energy by fish, chicken, and cattle

	Feed Composition			Efficiency		
Animal	Protein (%)	Energy (kcal ME/g)	ME–protein ratio (kcal/g)	Weight Gain per g of Food Consumed (g)	Protein Gain per g Protein Consumed (g)	Protein Gain per Mcal ME Consumed (g)
Channel catfish	32	2.7[a]	8.5	0.84	0.36	47
Broiler chicken	18	2.8	16.0	0.48	0.33	23
Beef cattle	11	2.6	24.0	0.13	0.15	6

[a] Metabolizable energy (ME) estimated from digestible energy.

Source: Adapted from Lovell (1979) and National Research Council (1983).

Table 11-2 Dressing percentage and carcass characteristics of various food animals

Source of Flesh	Dressing Percentage[a]	Characteristic of Dressed Carcass			Food Energy (kcal per 100 g of Edible Tissue)
		Refuse[b] (%)	Lean (%)	Fat (%)	
Channel catfish	60	14	81	5	112
Beef	61	15	60	25	147
Pork	72	21	54	26	147
Chicken	72	30	65	9	115

[a] The marketable percentage of the animal after slaughter.

[b] In fish, bones only; in beef and pork, bones, trim fat, and tendons; in poultry, bones only.

Sources: Channel catfish: Lovell (1979). Beef: Browning et al., (1988) and United States Department of Agriculture (1986). Pork: Prince, et al. (1987) and United States Department of Agriculture (1983). Poultry: Lovell (1988).

more than 80% of the dressed carcass of channel catfish is lean tissue; only 13.7% is bone, tendon, and waste fat. The caloric value of dressed fish is less than that of the edible protein of beef or pork. The net protein utilization (NPU) value of fish flesh, 83 (as compared to 100 for egg), is about the same as that of red meat, 80, although the essential amino acid profiles of fish and red meat both reflect high protein quality (Lovell, 1988).

The processing impact on the nutritional value of seafood protein is of concern. The protein quality of seafoods can be determined by either chemical and in vitro or bioassay methods. Acton and Rudd (1987) have recently reviewed various protein-quality methods for seafood. Therefore, the details of the methods will not be further discussed here. There are some problems when determining the quality of food proteins using animal experiments. These include length of time, high cost, and nonreproducibility. Additional problems such as fat oxidation, nonenzymatic browning, and deterioration caused by microorganisms occur in assessing the nutritional seafood protein quality using in vivo experiments. Because traditional methods for evaluating protein quality are based on meat and dairy proteins, they are often not appropriate for seafood proteins which have a much weaker structure and quite different compositional characteristics. Lee and Ryu (1987) has proposed c-PER assays for seafood protein-quality evaluation.

FISH MEAL AS PROTEIN SOURCE IN ANIMAL FEED

Introduction

Fish meal is used widely in animal feed as a protein source. Global use of fish meal is currently around 6.5 million tons (Pike, 1989). Consumption is now greatest in the Far East, followed by Western Europe, Eastern Europe, and by North and South America. Fish meal consumption in terms of species on a global basis has been estimated (Fig. 11–1) (Pike, 1989). Poultry are the biggest consumers followed by pigs and farmed fish. It is interesting to note that 7 years ago consumption by farmed fish was only 1–2% of global fish meal production. Five years later it was 10%.

Poultry

Continuing improvements are expected in the rate of growth of broilers and their feed conversion. Performance graphs are updated daily and can be compared with previous batches of boilers. Such man-

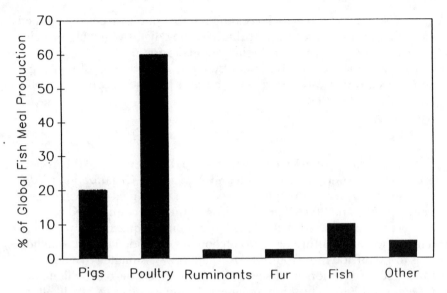

Figure 11–1   Fish meal consumption by species (global estimate for 1988).

Source: From Pike (1989).

agement control is leading to very rapid improvements in all aspects of production, especially diet formulation.

The nutrient concentration of diets is likely to increase further to achieve the production improvement by using higher values on amino acids and the higher-energy ingredients such as fish meal.

With increasing interest in ways of including the long-chain polyunsaturated $n-3$ fatty acids, especially eicosapentaenoic acid + docosahexaenoic acid, into the human diet, more interest will be focused on the inclusion of fish meal in broilers diet.

Pigs

Pig performance is also expected to continue to improve. As with poultry, nutrient concentration in feeds is likely to rise. The need to reduce waste by improving diet digestibility and phosphorus and nitrogen retention may favor the use of fish meal.

The amino acids in fish meal are more digestible than those in most vegetable proteins, including soybean meal. As this is taken into account in least-cost formation, it is likely to give a higher value for fish meal relative to vegetable proteins than present formulation techniques based on the content of total amino acids.

Another fact is that an increasing number of pigs are being weaned early, under 4 weeks of age. At this age, they digest vegetable proteins poorly. These proteins are highly antigenic for the very young pig, causing an inflammatory reaction of the gut wall. Because of their high cost, milk proteins are being replaced and fish meals offer a suitable alternative (Pike, 1989).

Fish

The farming of fish has grown very rapidly recently, and this growth is expected to continue. The Food and Agriculture Organization (FAO) in Rome has forecast that global production of farmed fish will double between 1986 and 2000. The expansion of the aquaculture feed production actually paralleled that of the expansion of aquaculture products. The world's aquaculture feed production has expanded from 1.9 million tons to 4.0 million tons from 1980 to 1988. It was estimated that aquaculture feed production in the year of 2000 would be 6.6 million tons (Meggison 1990). The higher-value fish and crustacean are likely to increase at more than the average rate.

Fish, like other animals, do not have a true protein requirement per se but have a requirement for a well-balanced mixture of essential and nonessential amino acids. However, fish nutrition seems quite different from that of on-land farm animals. Many fish require the same 10 essential amino acids. These are arginine, histidine, isoleucine, leucine, lysine, methionine, phenylalanine, threonine, tryptophan, and valine. Quantitative requirements for these amino acids are determined by feeding graded levels of one amino acid at a time in a test diet containing crystalline amino acids or a combination of a purified protein and crystalline amino acids. The amino acid profile of the test diet is usually similar to that in chicken or fish eggs or of the fish muscle. The complete quantitative amino acid requirements have been established for only five species, namely, the chinook salmon, common carp, Japanese eel, channel catfish, and tilapia (Table 11–3).

The minimum amount of dietary protein needed to supply adequate amino acids and produce maximum growth has been determined in many species and have been reviewed (Millikin, 1982; National Research Council, 1983). In general, the values range from about 30 to 55% crude protein of the diet for maximum growth. The values of protein requirement for fish are probably overestimated to some extent due to one or more of the following reasons: Most requirement values are based on levels of protein that result in maximum growth, with little or no index of protein utilization; many of the requirement values have

Table 11-3 Essential amino acids requirements of fish, chicken, and swine (percentage of the protein)

Amino Acid	Japanese Eel[a]	Common Carp[b]	Channel Catfish[b]	Chinook Salmon[b]	Tilapia nilotica[c]	Chicken[b]	Swine[b]
Arginine	4.2	4.2	4.3	6.0	4.2	5.6	1.2
Histidine	2.1	2.1	1.5	1.8	1.7	1.4	1.2
Isoleucine	4.1	2.3	2.6	2.2	3.1	3.3	3.4
Leucine	5.4	3.4	3.5	3.9	3.4	5.6	3.7
Lysine	5.3	5.7	5.1	5.0	5.1	4.7	4.4
Methionine	3.2	—	—	—	—	—	—
(+ cystine)	5.0	3.1	2.3	4.0	3.2	3.3	2.3
Phenylalanine	5.6	—	—	—	—	—	—
(+ tyrosine)	8.4	6.5	5.0	5.1	5.7	5.6	4.4
Threonine	4.1	3.9	2.0	2.2	3.6	3.1	2.8
Tryptophan	1.0	0.8	0.5	0.5	1.0	0.9	0.8
Valine	4.1	3.6	3.0	3.2	2.8	3.4	3.2

Sources: (a) from Arai (1986). (b) from National Research Council (1979, 1981, 1983, 1984). (c) from Santiago (1985).

been determined with varying protein-to-energy ratios; various investigators have used different estimated metabolizable energy values in formulating their test diets; and the protein sources may not contain adequate amounts of each of the essential amino acids (Wilson, 1985). The protein requirement values in fish are much higher than the on-land farm animals, in which less than 20% is often an optimum requirement. Consequently, only high-protein feedstuffs are included in fish feeds. Fish meal has traditionally been used as a major protein source because of its high nutritive value and palatability. A large number of plant protein feedstuffs has been used in domestic animal feeds, but relatively few are used in fish feeds because of the high dietary protein requirements. Therefore, only high-protein plantstuffs such as oilseed residues of soybean meal, peanut meal, cottonseed meal, sunflower seed meal, and rapeseed meal are used in fish feed. Of these, soybean meal appeared to be better utilized by most fish species as compared to other plant protein sources.

The replacement of fish meal components of fish feeds with soybean meal has been studied for quite some time. However, fish growth has almost always been retarded in direct relation to the level of replacement. This lack of success has led to the formulation of a number of hypotheses to explain the detrimental effects of substituting fish meal with soybean meal. These include (1) the presence of residual levels of trypsin inhibitors due to inadequate heating during soybean processing and (2) a less-than-optimal amino acid balance in soybean meal protein.

To date, fish meal still constitutes a substantial part of the feed formula as the protein source for aquacultured species.

Ruminants

The use of fish meal in ruminant diets—dairy cows, beef cattle, and sheep—has been established in Scandinavian countries and the United Kingdom. In other countries, its use for this purpose has been slow to develop.

FISH SILAGE AS PROTEIN SOURCE IN ANIMAL FEED

The use of fish meal as a major ingredient in commercial fish feeds is a common practice, despite extensive research into the possibility of replacing it with cheaper plant proteins. However, fish meal production is both capital and energy intensive as it requires 60–70 kg fuel/ton of raw fish (Windsor and Barlow, 1981), mainly due to the necessity of separating the raw material into the three fractions of oil, protein, and water. In contrast, the manufacture of commercial fish feeds often requires the recombining of fish protein with fish oil. The use of whole fish is, therefore, still widely used on aquaculture farms, despite the problems of continuity in supply and the expense of frozen storage.

Fish silage is a liquid product that can be prepared from whole fish or parts of fish (fish wastes) that are liquefied by the action of the endogenous gut enzymes of the fish in the presence of an added acid (Tatterson, 1982). The action of the enzymes breaks down the fish protein into smaller units of increased solubility, whereas the addition of acid increases the activity of the enzymes and also lowers the pH of the final material, thus inhibiting the growth of spoilage organisms and thereby ensuring the stability of the silage. The process is neither capital nor energy intensive and the product possesses good storage characteristics if correctly treated (Jackson et al., 1984). Moreover, it is widely used as a dietary ingredient for poultry and pigs (Raa and Gildberg, 1982). Although the potential for including it in fish feeds has been recognized (Asgard and Austreng, 1981), few formal studies have been conducted to assess its suitability as a dietary ingredient.

REFERENCES

ACTON, J. C., and RUDD, C. L. 1987. "Protein Quality Methods for Seafoods." Pp. 453–472. In *Seafood Quality Determination*, edited by Donald E. Kramer and John Liston, Amsterdam: Elsevier.

ARAI, S. 1986. Personal communication. National Research Institute of Aquaculture, Tamaki, Mie, Japan.

ASGARD, T., and AUSTRENG, E. 1981. "Fish Silage for Salmonids: A Cheap Way of Utilizing Waste as Feed." *Feedstuffs* 53(27):22–24.

BROWNING, M. A., HUFFMAN, D. L., EGBERT, W. R., and JONES, W. R. 1988. *Proceedings of Reciprocal Meat Conference*, Chicago, IL.

JACKSON, A. J., KERR, A. K., and COWEY, C. B. 1984. "Fish Silage as a Dietary Ingredient for Salmon. I. Nutritional and Storage Characteristics." *Aquaculture* 38:211–220.

LEE, K. H., and RYU, H. S. 1987. "Evaluation of Seafood Protein Quality as Predicted by c-PER Assays." Pp. 473–485. In *Seafood Quality Determination*, edited by Donald E. Kramer and John Liston. Amsterdam: Elsevier.

LOVELL, R. T. 1979. "Fish Culture in the United States." *Science* 206:1368–1372.

LOVELL, R. T. 1988. *Nutrition and Feeding of Fish*. New York: Van Nostrand Reinhold.

LOVELL, R. T., SHELL, E. W., and SMITHERMAN, R. O. 1978. *Progress and Prospects in Fish Farming*. Pp. 262–290. New York: Academic Press.

MEGGISON, P. A. 1991. "Future World of Aquaculture." In *Rovithai International Shrimp Seminar* July, 1990, Bangkok, Thailand.

MILLIKIN, M. R. 1982. "Effects of Dietary Protein Concentration on Growth, Feed Efficiency and Body Composition of Age–0 Striped Bass." *Trans. Amer. Fish. Soc.* 111:373–378.

NATIONAL RESEARCH COUNCIL. 1979. *Nutrient Requirements of Swine.* Washington, DC: National Academy of Sciences.

NATIONAL RESEARCH COUNCIL. 1981. *Nutrient Requirements of Coldwater Fish.* Washington, DC: National Academy of Sciences.

NATIONAL RESEARCH COUNCIL. 1983. *Nutrient Requirements of Warmwater Fish.* Washington, DC: National Academy of Sciences.

NATIONAL RESEARCH COUNCIL. 1984. *Nutrient Requirements of Poultry.* Washington, DC: National Academy of Sciences.

PIKE, I. H. 1989. "Global Use of Fish Meal and Developments Leading to Changes in Future Use." Pp. 10–22. In *Proceedings of Twenty-ninth Annual Conference.* International Association of Fish Meal Manufacturers.

PRINCE, T. J., HUFFMAN, D. L., BROWN, P. M., and GILLESPIE, J. R. 1987. Effects of Ractopamine on Growth and Carcass Composition of Finishing Swine." *J. Anim. Sci.* 65:309. (Suppl. 1).

RAA, J., and GILDBERG, A. 1982. "Fish Silage: A Review. *Crit. Rev. Food Sci. Nutr.* April:383–419.

SANTIAGO, C. B. 1985. *Amino Acid Requirements of Nile Tilapia.* Ph.D. dissertation, Auburn University, Auburn, AL.

TATTERSON, I. N. 1982. "Fish Silage: Preparation, Properties and Uses." *Anim. Feed Sci. Technol.* 7:153–159.

UNITED STATES DEPARTMENT OF AGRICULTURE. 1983. *Composition of Foods: Pork Products.* P. 23. Agriculture Handbook No. 8–10. Washington, DC: U. S. GPO.

UNITED STATES DEPARTMENT OF AGRICULTURE. 1986. *Composition of Foods: Beef Products.* P. 41. Agriculture Handbook No. 8–13. Washington, DC: U. S. GPO.

WILSON, R. P. 1985. "Amino Acid and Protein Requirements of Fish." Pp. 1–16. In *Nutrition and Feeding in Fish,* edited by C. B. Cowey, A. M. Mackie, and J. G. Bell. London: Academic Press.

WINDSOR, M., and BARLOW, S. 1981. *Introduction to Fishery By-Products.* Farnham, Surrey: Fishing News Books.

YOUNG, V. R., and PELLET, P. L. 1987. "Protein Intake and Requirements with Reference to Diet and Health." *Amer. J. Clin. Nutr.* 45:1323–1343.

12

PROTEINS FROM SEAFOOD PROCESSING DISCARDS

Fereidoon Shahidi

INTRODUCTION

Processing discards from fisheries account for as much as 70–85% of the total weight of catch and these have been generally dumped in-land or hauled into the ocean. Discards have typically included processing wastes from fish, shellfish, and by-catch of unutilized species as well as underutilized species such as male capelin and seal carcasses. However, environmental restrictions have brought about a pressing need for devising novel methods for utilization of processing by-products and discards.

Composting of fisheries waste provides a means for their so-called "green" disposal. Preparation of silage and fish meal also present ways for using proteins in animal and aquaculture feed formulations. However, the main area of progress has been in preparation and utilization of value-added products. These include preparation of surimi from fish scraps and frames, fish protein concentrates, low-temperature enzymes, antifreeze proteins and flavorants from shellfish discards, and gelatin and glue from skin and frames of aquatic species as well as other biologically active components.

Utilization of value-added components extracted from seafood processing by-products in a variety of applications has been practiced. In particular, use of antifreeze proteins for extending the shelflife of human organs intended for transplant purposes, use of low-temperature enzymes as processing aids in the preparation of a variety of food products, use of chitin and chitosan in food, agriculture, pharmaceuticals, waste treatment, and water purification may be noted. Utilization of carotenoid extracts from shellfish processing discards in aquaculture feed for salmonids and use of extracted flavorants in the preparation of shellfish analogs may also be important. In addition, marine lipids have received considerable attention in recent years as both food and feed components as well as pharmaceutical commodities. This chapter provides information on proteins of seafood processing discards.

CLASSIFICATION OF FISHERIES WASTES

Fisheries wastes may originate from (1) processing discards of existing catches such as those of ground-fish and shellfish; (2) unutilized marine species and/or stocks that are not fished intentionally; and (3) underutilized species which have directed fisheries but may be considered as surplus.

Fisheries processing by-products are currently (1) dumped in-land or in the oceans, (2) composted or used as fertilizer, (3) used in the preparation of meal and oil, (4) used for silage preparation, and (5) used for the preparation of value-added components. Although value-added utilization of fisheries wastes is desirable, economic viability of products and production costs must be realistically assessed. Therefore, enforcement of environmental restrictions and higher costs associated with dumping of discards together with appropriate economic return from further processing of waste and novel approaches and designs for utilization of products is of prime importance when evaluating such processes.

In general, it is desirable to combine primary processing of fish and shellfish and utilization of by-products with aquaculture operations. Processing discards generally contain proteins, lipids, pigments, and enzymes, as well as other components which could be readily used as components of feed for cultured species. Figure 12-1 illustrates the integration scheme for fish processing and aquaculture.

FISH PROCESSING DISCARDS
AND UNUTILIZED/UNDERUTILIZED SPECIES

Processing discards account for up to three-quarters of the total weight of fish. Only about 25% of the total weight of fish is used as fillets (Piggott,

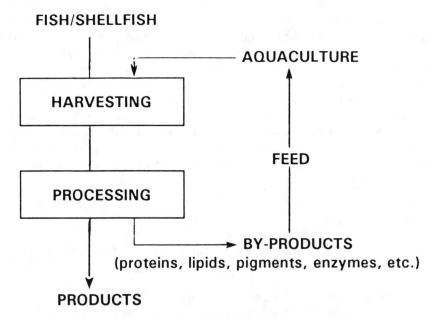

FISH/SHELLFISH

HARVESTING

AQUACULTURE

FEED

PROCESSING

BY-PRODUCTS
(proteins, lipids, pigments, enzymes, etc.)

PRODUCTS

Figure 12–1 Scheme for integrated fish/shellfish processing and aquaculture operations.

1986). Another 25% of mince could be obtained using mechanical deboning procedures. The resultant minced meat may be used for the preparation of fish cakes or surimi intended for kamaboko production. In the case of herring and lumpfish, and more recently capelin, roe from the female are extracted and the carcass is usually dumped. Furthermore, male and spent capelin are also discarded. Seal carcasses after removal of pelt, blubber, and flippers are dumped if no demand for their further processing and utilization is in sight.

Table 12–1 indicates that capelin and mechanically separated cod meat samples have similar compositional characteristics. However, seal

Table 12–1 Proximate composition of selected seafoods/discards

Component (%)	Capelin	Seal Meat	Cod	Cod Offal
Moisture	78.10 ± 0.08	70.84 ± 0.11	81.22 ± 0.04	77.61 ± 0.35
Crude protein	13.90 ± 0.71	23.21 ± 0.13	17.81 ± 0.04	14.30 ± 0.61
Lipid	3.56 ± 0.27	4.30 ± 0.78	0.67 ± 0.01	3.69 ± 0.02
Ash	2.41 ± 0.02	3.95 ± 0.25	1.16 ± 0.01	1.97 ± 0.02

Source: Adapted, in parts, from Shahidi et al. (1991) and Shahidi et al. (1992c).

meat had a much higher and cod offal a somewhat lower content of proteins. Meanwhile, the latter had a lower quality of proteins as reflected in the calculated protein efficiency ratio (PER) values (Shahidi et al., 1991) obtained by an amino acid scoring procedure developed by Lee et al. (1978).

The lower protein quality of cod offal as compared with other muscle proteins was caused by the presence of a large proportion of connective tissues and skin in the total crude proteins of the product (Table 12–2). The presence of a large amount of collagen in the discards brought about an increase in their content of glycine proline, and hydroxyproline.

SURIMI

Surimi is the mechanically deboned fish flesh that is repeatedly washed in 5–10°C water until most or all of its water-soluble proteins are removed

Table 12–2   Amino acid composition of capelin, seal meat, cod, and cod offal as percentage of total proteins

Amino Acid/PER Value	Capelin	Seal Meat	Cod	Cod Offal
Alanine	5.57	5.88	4.80	6.82
Arginine	5.99	6.21	5.99	6.50
Aspartic acid	8.86	8.23	10.24	9.05
Cysteine	1.33	0.87	1.07	0.85
Glutamic acid	13.19	11.46	14.92	12.56
Glycine	5.32	4.47	6.05	11.72
Histidine	2.43	5.01	2.94	1.79
Hydroxyproline	0.42	0.55	0.38	2.38
Isoleucine	4.72	4.58	4.61	3.30
Leucine	8.15	7.44	8.13	6.18
Lysine	8.47	8.92	9.18	6.53
Methionine	3.09	1.64	2.96	2.78
Phenylalanine	3.80	4.57	3.90	3.18
Proline	3.70	3.89	3.54	5.57
Serine	4.18	3.98	4.08	5.35
Threonine	4.28	4.53	4.39	4.15
Tryptophan	1.07	1.20	1.12	1.01
Tryosine	3.34	2.85	3.38	2.36
Valine	5.71	5.80	5.15	3.80
PER value	2.86–2.98	2.90–2.99	2.90–2.99	2.31–2.33

Source: Adapted, in parts, from Shahidi et al. (1991) and Shahidi et al. (1992c).

and to which a cryoprotectant mixture is added (see Chapter 10; also Lee, 1984). Of particular interest to the present chapter is the preparation of surimi from fish frames and from unutilized/underutilized fish species and seal meat.

Washing of minced fish with water generally removes water-soluble proteins (albumins) as well as enzymes and blood residues. The amino acid profile of seal surimi prepared by washing the mechanically separated seal meat (MSSM) with aqueous solutions is similar to that of the original meat. However, the content of glycine, proline, and hydroxyproline, all dominant amino acids of collagen, is increased. Meanwhile, almost all of the taurine present in MSSM is removed after aqueous washings, perhaps because this compound existed in the free form (Table 12–3). Furthermore, most of the low-molecular-weight compounds with potential flavor-activity characteristics such as free amino acids (Tables

Table 12–3 Amino acid composition of raw and washed mechanically separated seal meat (MSSM) as percentage of total proteins

| Amino Acid/PER Value | MSSM | Washed MSSM | |
		$2\times H_2O$	$1\times H_2O$, $1\times 0.5\%$ $NaHCO_3$
Alanine	5.88	6.05	5.58
Arginine	6.21	7.06	6.20
Aspartic acid	8.23	8.55	8.80
Cysteine	0.87	0.95	1.49
Glutamic acid	11.46	11.33	13.47
Glycine	4.47	6.17	5.47
Histidine	5.01	2.88	3.31
Hydroxylysine	0.10	0.40	0.18
Hydroxyproline	0.55	1.73	1.24
Isoleucine	4.58	4.59	4.83
Leucine	7.44	6.33	8.15
Lysine	8.72	7.82	8.50
Methionine	1.64	1.49	1.93
Phenylalanine	4.57	4.09	4.00
Proline	3.89	5.24	4.12
Serine	3.98	3.72	3.58
Threonine	4.53	3.83	4.00
Tryptophan	1.20	1.06	1.05
Tryosine	2.85	2.90	3.07
Valine	5.80	5.46	5.04
PER Value	2.67–2.99	2.4–2.67	2.82–2.84

Source: Adopted from Shahidi and Synowiecki (1992).

Table 12–4 Free amino acid content (mg/100 sample) in mechanically
separated seal meat (MSSM) and washed MSSM

Amino Acid	MSSM	Washed MSSM	
		$2\times H_2O$	$1\times H_2O$, $1\times0.5\%$ NaHCO$_3$
Alanine	41.43	4.63	3.77
β-Alanine	2.07	—	—
α-Aminobutyric acid	2.23	—	—
Arginine	17.22	4.08	3.38
Aspartic acid	9.07	2.17	1.71
Asparagine	3.12	0.42	0.17
Citruline	1.61	0.08	0.04
Cysteine	1.05	0.48	0.36
Glutamic acid	20.28	3.79	3.32
Glutamine	32.74	3.36	2.86
Glycine	13.96	2.19	1.69
Histidine	9.05	2.05	1.39
Hydroxyproline	2.98	0.15	—
Isoleucine	8.93	1.39	1.09
Leucine	20.75	3.86	2.51
Lysine	18.80	4.38	3.58
Methionine	9.16	1.96	1.57
Methylhistidines (1 & 3)	0.36	0.08	0.08
Ornithine	2.15	0.33	0.25
Phenylalanine	10.20	1.89	1.29
Proline	9.60	1.47	1.06
Serine	17.09	3.17	2.37
Taurine	59.34	7.64	5.79
Threonine	12.93	2.19	1.57
Tryptophan	1.75	0.49	0.30
Tryosine	9.86	1.96	1.46
Valine	15.42	2.79	1.85

Source: Adapted from Shahidi et al. (1992c).

12–4 and 12–5) and nucleic acids and amines (Table 12–5) as well as
minerals and vitamins (Table 12–6) are removed during the washing of
seal meat.

SEAFOOD PROTEIN PREPARATIONS

Seafood protein preparations encompass protein concentrates/hydroly-
zates, meals, and silage from unutilized/underutilized species and pro-

Table 12–5   Contents (mg N/100 g sample) of some nitrogen-containing compounds in mechanically separated seal meat (MSSM) and washed MSSM

Component	MSSM	Washed MSSM	
		$2 \times H_2O$	$1 \times H_2O$, $1 \times 0.5\%$ NaHCO$_3$
Free amino acids	51.87	8.53	6.03
Amines	1.74	0.78	0.62
Ammonia	5.10	0.81	0.71
Nucleic acids	40.73	23.10	19.90
Imidazole dipeptides	94.87	11.56	9.50
Total NPN	341.66	57.51	36.38

Source: Adapted from Shahidi et al. (1992c).

Table 12–6   Effects of washing on the nutrient quality of mechanically separated seal meat (MSSM)

Component (in 100 g Sample)	Dimension	MSSM	Washed MSSM	
			$2 \times H_2O$	$1 \times H_2O$, $1 \times 0.5\%$ NaHCO$_3$
Total crude protein	g	23.21	13.18	12.17
Collagen	g	0.92	1.65	1.09
PER value	—	2.99	2.67	2.81
Lipid	g	3.69	1.61	1.15
Calorie value	kJ	5.28	2.82	2.47
Calcium	mg	591.04	1330.01	991.15
Phosphorus	mg	504.05	560.15	565.00
Potassium	mg	288.00	24.50	22.20
Sodium	mg	159.00	36.40	174.00
Iron	mg	164.60	28.90	11.90
Magnesium	mg	34.20	33.30	21.91
Zinc	mg	2.80	3.00	2.05
Thiamin	mg	0.11	0.02	0.02
Riboflavin	mg	0.35	0.23	0.10
Niacin	mg	6.10	0.40	0.40
Pyridoxine	mg	0.23	0.06	0.02
Vitamin B12	μg	7.70	4.50	2.40
Pantothenic acid	mg	0.89	0.29	0.19
Folic acid	μg	3.30	2.50	1.77

Source: Adapted from Shahidi and Synowiecki (1992).

cessing by-products. Depending on the starting material, the characteristics, quality, and, hence, utilization of products varies. However, all products may be used in either human food formulations or in the preparation of animal feed.

Fish Protein Concentrates/Hydrolyzates

Preparation of products from both lean and fatty fish species is possible. Lean fish generally affords the so-called "type A" concentrate which contains at least 67.5% protein and not more than 0.75% lipid. Products of this type are generally colorless, tasteless, and odorless. Concentrates from fatty fish provide the "type B" product with a lipid content of up to 10% with a distinct fishy flavor. Both products may be used as food components; however, type B may be more suitable for populations accustomed to fish-containing diets (Ockerman and Hansen, 1988).

Protein concentrates from marine resources with a bland taste and an ivory-white color have been prepared using proteolytic enzymes. Proteolytic enzymes of commercial origin as well as natural enzymes isolated from fish intestine were employed to prepare high-quality products from capelin (Shahidi et al., 1992a), seal meat, and seal bone residues (Shahidi et al. 1992b). Meanwhile, modification of fish proteins in this manner followed by separation of protein–phosphate complexes using sodium hexametaphosphate and subsequent precipitation of the complex by lowering the pH to 2.5–3.5 has been attempted (Ockerman and Hansen, 1988). The precipitated complex is then centrifuged, washed, and solvent extracted to remove lipids. The products may be neutralized with a base and mixed with 10–20% carbohydrates and dehydrated by drum-, spray-, or freeze-drying. Figure 12–2 shows a typical flowsheet for preparation of a hydrolyzate from marine resources.

The taste of products obtained by proteolytic modification of proteins depends on their degree of hydrolysis. Extensive hydrolysis may lead to the formation of bitter peptides. Therefore, controlled hydrolysis with appropriate enzymes for different products is recommended (Figure 12–3).

The amino acid compositions of hydrolyzates from both seal and capelin were generally similar to those of their starting material (Table 12–7). The products so obtained may possess moderate antioxidant activity, perhaps due to their ability to chelate metal ions (unpublished data).

Fish Meal and Silage

Fish meal is an excellent source of feed for animal nutrition. Its content of essential amino acids, B vitamins, and minerals including

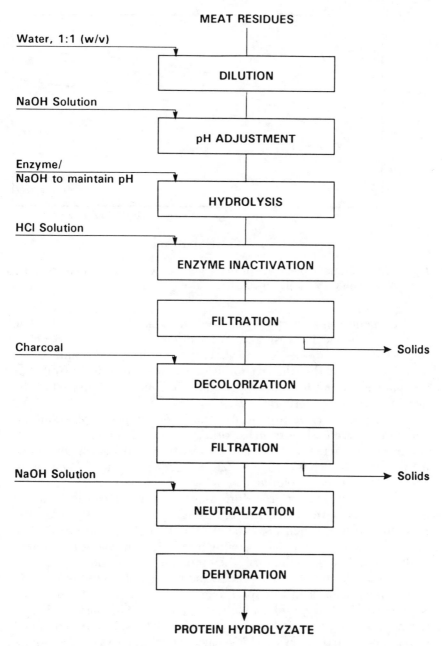

Figure 12–2 Preparation of hydrolyzate for meat residues on fish frame or animal carcass bones using an alkaline active enzyme.

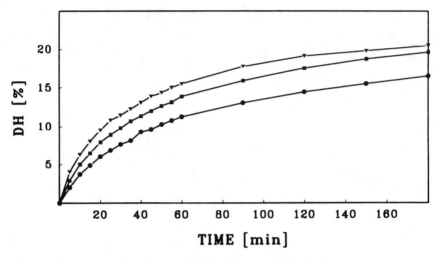

Figure 12–3 Time dependence of the degree of hydrolysis (DH) at 50°C (●), 55°C (■), and 60°C (▼), using an alkaline active enzyme.

trace elements is high and well balanced (Karrick, 1976; Lee 1976). The production of meal is easy and its inclusion in feed rations depends somewhat on the species of animal. To prevent fishy odors and flavors in the products, either the level of fish meal in the feed is controlled or the animals are put for a fixed period of time on rations devoid of the meal prior to slaughtering.

Fish meal may be produced from different species of unutilized and underutilized fish or from fish offal, scraps, and cannery wastes. The raw material is generally cooked and pressed to yield 12–18% products, nearly 75% of which is the meal. Both wet reduction and dry rendering procedures may be employed.

The wet reduction procedure uses a continuous cooker in which the material is moved by a conveyer through the cooker to coagulate the proteins which are then dehydrated in a continuous screwpress. The press cake is flaked and subsequently dried to a final moisture content of 8% (w/w). The press liquor is then separated to oil and stickwater.

Dry rendering is a batch process and is primarily used for preparation of meal from lean fish species and discards as well as seal which is properly dressed (all of its blubber removed). The material is first comminuted and then fed to a steam-jacketed cooker–dryer equipped with a power-stirring device which may operate under vacuum or atmospheric pressure. The meal is finally pressed and dehydrated. In all cases, the products are stabilized by the addition of 200 ppm of ethoxyquine, on a dry weight basis.

Table 12–7 Amino acid composition of protein hydrolyzate from seal meat and capelin obtained using Alcalase

Amino Acid	Seal Meat	Seal Protein Hydrolyzate	Capelin	Capelin Protein Hydrolyzate
Alanine	5.88	5.80	5.57	6.00
Arginine	6.21	6.02	5.99	5.70
Aspartic acid	8.23	8.90	8.86	9.89
Cysteine	0.87	1.01	1.33	1.34
Glutamic acid	11.46	12.53	13.19	13.43
Glycine	4.47	5.58	5.32	5.14
Histidine	5.01	5.25	2.43	2.09
Hydroxylysine	0.04	0.10	0.	0.17
Hydroxyproline	0.75	0.85	0.42	0.46
Isoleucine	4.58	3.92	4.72	4.25
Leucine	7.44	8.50	8.15	7.60
Lysine	8.72	9.44	8.47	8.49
Methionine	1.64	1.62	3.09	2.05
Phenylalanine	4.57	3.96	3.80	3.19
Proline	3.89	3.86	3.70	3.67
Serine	3.98	3.60	4.18	4.24
Threonine	4.53	3.87	4.28	4.56
Tryptophan	1.20	1.19	1.07	0.43
Tyrosine	2.85	2.46	3.34	2.47
Valine	5.80	4.62	5.71	5.77
PER value	2.67–2.90	2.80–2.90	2.86–2.98	2.64–2.79

Note: The results presented are unpublished.

Fish silage is a liquid product prepared mostly from fish waste and from underutilized/unutilized species of fish using organic or inorganic acids, usually formic acid, which creates a suitable environment for the endogenous enzymes, mostly pepsin, to degrade the tissues (see Chapter 9; also Raa and Gildberg, 1982; Gildberg and Almås, 1986). The process has been used commercially for sometime for production of animal and aquaculture feed. Acid digestion of fish at low-salt concentrations followed by neutralization and salt addition may be used as a means to accelerate fish sauce production (Gildberg et al., 1984).

ANTIFREEZE PROTEINS AND ANTIFREEZE GLYCOPROTEINS FROM FISH BLOOD

Antifreeze proteins (AFP) and antifreeze glycoproteins (AFGP) are found in the blood of several species of fish in polar and north temperate oceans. These are synthesized in the liver and exported to blood where they lower the plasma freezing point to $-1.9°C$ as compared with $-0.7°C$ for unprotected plasma and, therefore, protect tissues from damage at hypothermic temperatures. Although the AFGP from different fish species are similar in their chemical structures, there are three broadly classified AFP referred to as types I, II, and III (Rubinsky et al., 1991). Both AFGP (3–26kDa) and AFP (3.3–14 kDa) are generally small in size and may be separated from the bulk of other plasma proteins by two cycles of gel permeation chromatography followed by reversed-phase HPLC (Hew and Fletcher, 1985; Hew and Davis, 1987).

AFP of type I were first identified in the northern winter flounder (*Pseudo pleuronectes americanus*). These AFP, found in sculpins and flounders, consist of a family of seven independently active compounds which are alanine rich (60 mol%) ranging in molecular weight from 3.3 to 4.5 kDa with amphiphilic α-helices. The type II AFP is larger than type I (14 kDa), has a β structure and five disufide bridges. This AFP has been isolated from the sea raven (*Hemitripterus americanus*), Atlantic herring, and smelt. The type III AFP was isolated from Newfoundland ocean (eel) pout (*Macrozoarces americanus*) and have a molecular weight of 5.0–6.7 kDa, are not alanine rich, contain no cysteine residues, and have no distinguishing features (Table 12–8). Similar AFP have been isolated from Antarctic zoarcids (Rubinsky et al., 1991).

The use of antifreeze proteins, isolated from blood plasma, to protect mammalian calls at cryogenic temperatures ($-130°C$) has recently been reported (Rubinsky et al., 1991). The potential practical value of this finding is of paramount importance to the short-term preservation of cold-sensitive mammalian cells and human organs intended for transplant purposes. Furthermore, using some combination of improved promoters and increased gene dosage, it is likely that circulating AFP levels can be boosted to match those seen in nature. Therefore, due to the fact that AFP is functional in foreign hosts, there is a real possibility that a fully freeze-resistant fish can be produced by transgenic methods.

FISH GELATIN AND GLUE

Gelatin and glue are water-soluble, hydrophilic products produced by hydrolysis of the fibrous connective tissues of skin and bones. The quality of gelatin or glue is dictated by its molecular-weight distribution and other variables. Gelatin is used in food, pharmaceuticals, and photographic and research fields for production of jellies, candies, marshmallows, whipped creams, capsules, clarification of wines, photographic

Table 12–8 Characteristics of fish antifreeze proteins (mol%)

Amino Acid/ Structure	AFPI Winter Flounder	AFPI Sculpin	AFPII Sea Raven	AFPIII Ocean Pout	AFGP Cod
Aspartic acid	11.5	6.4	10.7	1.2	
Threonine	9.9	6.4	7.9	11.4	33
Serine	3.3	1.6	8.2	9.7	
Proline	0.	1.6	6.7	8.9	+
Glutamic acid	1.6	3.2	9.1	12.4	
Glycine	0.	1.6	8.1	6.4	
Alanine	62.8	60.8	14.4	11.2	66
Half cystine	0.	0.	7.6	0.8	
Valine	0.	0.	1.2	11.	
Methionine	0.	2.4	5.4	1.2	
Isoleucine	0.	1.6	1.7	6.8	
Leucine	5.8	4.	6.2	4.7	
Tyrosine	0.	0.	1.2	1.3	
Phenylalanine	0.	0.	2.	1.5	
Lysine	3.3	8.	2.1	6.2	
Histidine	0.	0.	2.5	0.	
Tryptophan	0.	0.	2.8	ND	
Arginine	1.6	2.4	2.3	0.5	
Hexosamine	0.	0.	0.	0.	+
MW	4,000	3,000–5,000	13,000	6,500	2,600–33,000
Secondary Structure	α-Helix	α-Helix	β-Structure		Expanded structure

Source: Adapted from Davis et al. (1989).

films, and glues. Gelatin may retain more than 50 times its weight of water within its gel structures (Glicksman, 1969). Commercial gelatins contain approximately 88% proteins, 10% moisture, and 1–2% salts (Glicksman, 1969; Ward and Courts, 1977). Gelatin derived from the skin of deep-cold-water fish has a lower amount of proline and hydroxy-proline and, thus, gels at 8–10°C and not at room temperature (Table 12–9).

Fish gelatin, in addition to its common food uses, may be used in a light-sensitive coating and as a base for photoresist (Holahan, 1965). It is also used in the manufacture of lead frames that hold the silicon chip in computers and microprocessors. Its application in color video cameras has also been reported (Norland, 1990).

Table 12–9   Amino acids of lumpfish and commercial (Knox) gelatins as percentages of total proteins

Amino Acid	Knox	Lumpfish
Alanine	8.3	8.2
Arginine	7.4	7.7
Aspartic acid	4.9	7.3
Cysteine	0.3	0.1
Glycine	20.8	21.5
Glutamic acid	8.3	9.3
Histidine	0.7	0.9
Hydroxylysine	1.0	1.1
Hydroxyproline	9.4	6.5
Isoleucine	1.1	1.2
Leucine	2.7	2.5
Lysine	3.5	3.2
Methionine	1.0	0.9
Phenylalanine	2.0	2.5
Proline	14.5	9.5
Serine	2.4	6.7
Threonine	1.5	2.9
Tyrosine	0.7	0.7
Valine	2.4	2.5

Source: Adapted from Osborne et al. (1990).

FISH ENZYMES

Fish processing wastes provide large amounts of gut enzymes which have potential use in a variety of applications (Simpson and Haard, 1987). For example, 1 kg of cod stomach or intestine contains approximately 2 g of pepsin and 1 g of trypsin, chymotrypsin, and elastase (Gildberg, 1992). The enzymes present in fish guts are significantly different from their counterparts in warm-blooded animals and may offer distinct advantages for certain applications. For example, fish pepsins have been reported to have a higher pH optimum than other pepsins and are active at low temperatures. Furthermore, fish pepsins are resistant to autolysis at a low pH (Raa, 1990). Unique characteristics of fish enzymes are due to the existing differences in their amino acid composition with mammalian enzymes (Table 12–10) (Gildberg and Øverbø, 1990). In addition, the autolysis of whole fish during ensilation is mainly due to the presence of gut enzymes and, therefore, fillets alone poorly liquefy (Raa and Gildberg, 1976).

Application of fish enzymes as processing aids for preparation of

Table 12–10   Amino acid composition of some proteinase enzymes

Amino Acid	Number of Amino Acids in Enzymes			
	Cod Elastase	Porcine Elastase	Cod Trypsin	Bovine Trypsin
Aspartic acid	21	24	19	22
Threonine	21	19	10	10
Serine	33	22	23	34
Glutamic acid	10	19	27	15
Proline	9	7	6	8
Glycine	35	25	28	25
Alanine	16	17	16	14
Valine	26	27	18	15
Cysteine	9	8	12	12
Methionine	6	2	6	2
Isoleucine	12	10	8	13
Leucine	21	18	16	13
Tryosine	6	11	10	10
Phenylalanine	4	3	3	3
Tryptophan	6	7	3	3
Lysine	7	3	8	14
Histidine	10	6	6	3
Arginine	14	12	6	2
Total	266	240	225	217

Source: Adapted from Gildberg and Øverbø (1990).

caviar from a variety of fish species, for removal of squid skin, for cleaning of scallop, and for descaling of fish has been successfully achieved. In each case, the enzymes from fish viscera selectively break the linkage between the eggs, proteins, and so on and other connective tissues without damaging the desired end product (Raa, 1990). However, in some of these applications, commercial enzymes with mild proteolytic activity such as Neutrase may prove useful (unpublished results). Furthermore, use of gastric proteases from harp seal (*Phoca groenlandica*) for coagulation of milk has been reported (Shamsuzzaman and Haard, 1983). Cheddar cheese prepared with the aid of seal gastric proteases received higher sensory acceptability scores than that made with calf rennet.

The waste from fish processing may be used for the production of enzymes suitable for different commercial and biotechnological purposes. Several authors have reported the isolation and purification of fish enzymes such as pepsin from cod (Almås, 1990) and capelin (Gildberg and Raa, 1983), trypsin (Martinez et al., 1988) and cathepsin from the digestive glands of squid (Gildberg, 1987), and alkaline phosphatase from shrimp (Olsen et al., 1991). The latter may have use in the preparation of diagnostic kits.

Shellfish species such as shrimp contain enzymes such as chitinase and acetylglucosaminidase which are active in chitin metabolism. During moult, part of the chitin present in the old exoskeleton is recycled for synthesis of the new shell (Chen, 1987). This requires the presence of both chitin-degrading and chitin-synthesizing enzymes. Recovery of these enzymes from wastewater from shrimp processing industries, as well as hyaluronidase and alkaline phosphatase, has been described by Olsen et al. (1990). Recovery of these enzymes has been shown in the flowsheet of Figure 12–4.

HEMOPROTEINS AND CAROTENOPROTEINS

Hemoproteins, namely, hemoglobin and myoglobin, are responsible for transport of oxygen in the blood and tissues of animals, respectively. Carotenoproteins and carotenoids are other classes of compounds found in the flesh and skin of salmonid fish and in the exoskeleton of shellfish. Fish like other animals are unable to perform a de novo synthesis of carotenoids and these compounds are, therefore, deposited in fish tissues from dietary intakes (Haard, 1992).

Both hemoproteins and carotenoids are composed of a protein moiety and a nonprotein prosthetic group. The heme portion may be recovered from blood as well as muscles. The recovered material may be used as a source of readily absorbable iron or as a chemical substrate for production of the cooked cured-meat pigment (Shahidi et al., 1987). Isolation of hemin from animal blood and as a by-product in preparation of hydrolyzates from seal meat has been reported (Shahidi et al., 1992b). Isolation of carotenoproteins and carotenoids from shellfish processing discards has also appeared in the literature (Long and Haard, 1988; Shahidi and Synowiecki, 1991). Inclusion of these latter pigments in feed formulations intended for aquacultured salmonids has shown the importance of these compounds in industrial applications (Shahidi et al., 1993).

NONPROTEIN NITROGEN (NPN) COMPOUNDS

Seafood products contain a higher content of NPN compounds as compared to other muscle foods. The relatively high concentration of NPN compounds in seafoods presents unique, delicate, and different flavors

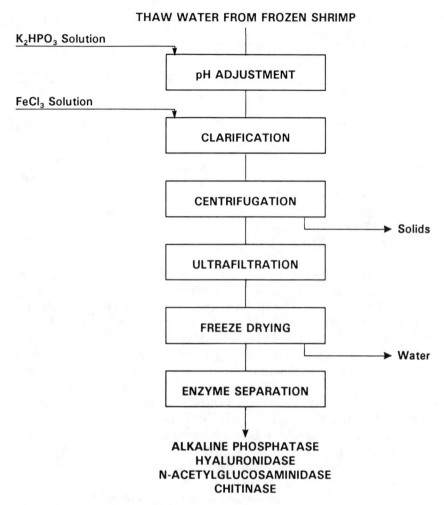

Figure 12–4 Recovery of enzymes from thaw water of frozen shrimps.

to products. However, some of these compounds are also responsible for the rapid deterioration of seafood quality because they serve as substrates for typical spoilage organisms that convert them to volatile bases with unfavorable odors.

Table 12–11 gives the percent distribution of NPN compounds in some seafoods. The free amino acids and nucleotides are perhaps the most important components contributing to the desirable flavor of seafoods. However, amines may be responsible for the production of undesirable odors in the stored seafoods.

Table 12–11 Percentage distribution of nonprotein nitrogen compounds in selected fish and shellfish

Species	Free Amino Acids	Nucleotides	Peptides	Creatine/ Creatinine	Betaines	Others[a]
Mackerel	25	10	5	35	—	25
Shark	5	5	5	10	—	75[b]
Shrimp	65	5	5	—	5	20
Crab	75	5	15	—	5	—
Squid	50	5	15	—	10	20[c]
Clam	35	5	35	—	15	10

[a]Ammonia, urea, TMAO, and amides.
[b]Primarily urea.
[c]Includes 10% octopine.
Source: Adapted from Finne (1992).

The free amino acids of seafoods are similar to those from other muscle foods. However, some unique compounds such as taurine, sarcosine, β-alanine, 1- and 3-methylhistidines, and α-aminobutyric acid may also be present (Table 12–4). Furthermore, the content of free glutamic acid, along with other free amino acids, is related to the taste of seafoods.

Fish with dark muscles have a generally high content of free histidine and a moderate amount of other free amino acids. However, mollusks, crustaceans, and echinoderms have characteristic distributions of amino acids which always include a certain amount of arginine, glycine, alanine, proline, and glutamic acid. Shrimp, lobster, crab, and other shellfish generally contain larger amounts of glutamic acid, glycine, and alanine than fish does. In addition, glycine betaine and certain dipeptides such as carnosine and anserine are said to contribute to the taste of seafoods (Komata, 1990).

Deamination or decarboxylation of free amino acids by spoilage microorganisms or by decarboxylase enzymes, respectively, may result in the formation of breakdown products. The content of ammonia produced by deamination of seafood constituents has been suggested to serve as an objective quality indicator for fresh crab and fish. Meanwhile, decarboxylation of amino acids such as arginine and lysine leads to the formation of polyamines putrescine and cadaverine, respectively. Decarboxylation of histidine results in the formation of histamine, the most potent capillary dilator which is a cause of food poisoning in certain seafoods such as tuna.

Nucleotides, primarily adenosine-5′-triphosphate and its monophosphate and diphosphate analogs are the dominant nucleotides in live tissues. Their enzymatic degradation during the postmortem period results in the formation of a number of products shown in Figure 12–5 (Terasaki et al., 1965). In most cases, the rate-determining step is (5) or (6), depending on the species of animal/fish under investigation (Karube and Sode, 1988). Consequently, the concentration of inosine and hypoxanthine increases with the length of storage period and these may be used as freshness indicators for seafoods.

The contribution of 5′-nucleotides to the taste of fish and shellfish is well recognized. The so-called umami taste of seafoods such as sea

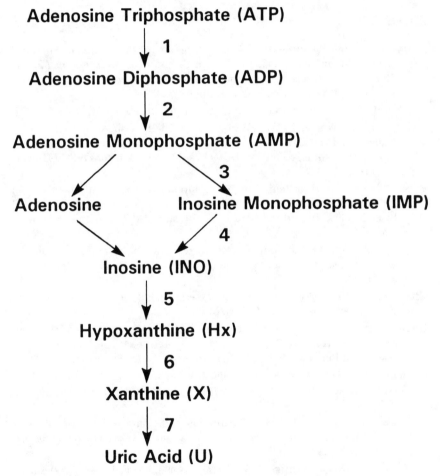

Figure 12–5 Degradation of adenosine–5′-triphosphate during the postmortem period.

urchins and abalone is attributed to nitrogenous compounds, mainly nucleotides (Komata, 1990). Therefore, low-molecular-weight nonprotein nitrogen compounds play an important role in the flavor quality of seafoods. The content of flavor-active ingredients isolated from shrimp and crab processing discards by a water-extraction process was approximately 1.0–1.5% of the total weight of protein (Shahidi and Synowiecki, 1991). The isolated flavorants from discards or from cook water in crab processing may be utilized further to prepare flavorants for use in the preparation of kamaboko products.

REFERENCES

ALMAS, K. A. 1990. "Utilization of Marine Biomass for Production of Microbial Growth Media and Biochemicals." In Pp. 361–373. *Advances in Fisheries Technology for Increased Profitability,* edited by M. N. Voigt and J. R. Botta. Lancaster, PA: Technomic Publishing Co.

CHEN, A. C. 1987. "Chitin Metabolism." *Arch. Insect. Biochem. Physiol.* 6:267–277.

DAVIS, P. L., FLETCHER, G. L., and HEW, C. L. 1989. "Fish Antifreeze Protein Genes and Their Use in Transgenic Studies." In Pp. 85–109. *Oxford Surveys on Eukaryotic Genes,* edited by N. MacLean, Vol. 6, Oxford: Oxford University Press.

FINNE, G. 1992. "Non-protein Nitrogen Compounds in Fish and Shellfish." In Pp. 393–401. *Advances in Seafood Biochemistry: Composition and Quality,* edited by G. L. Flick, Jr. and R. E. Martin. Lancaster, PA: Technomic Publishing Co.

GILDBERG, A. 1987. "Purification and Characterization of Cathepsin D from the Digestive Gland of the Pelagic Squid *Todarodes Sagittatus.*" *J. Sci. Food Agric.* 39:85–94.

GILDBERG, A. 1992. "Recovery of Proteinases and Protein Hydrolyzates from Fish Viscera." *Bioresource Technol.* 39:271–276.

GILDBERG, A., ALMAS, K. A. 1986. "Utilization of Fish Viscera." In Pp. 383–390. *Food Engineering and Process Application,* edited by M. Le Maguer and P. Jelen. London/New York: Elsevier Applied Science Publishers.

GILDBERG, A., and ØVERBØ, K. 1990. "Purification and Characterization of Pancreatic Elastase from Atlantic Cod (*Gadus morhua*)." *Comp. Biochem. Physiol.* 97B:775–782.

GILDBERG, A., and RAA, J. 1983. "Purification and Characterization of Pepsins from the Arctic Fish Capelin (*Mallotus villosus*)." *Comp. Biochem. Physiol.* 75A:337–342.

GILDBERG, A., ESPEJO-HERMES, J., and MAGAS-OREJANA, F. 1984. "Acceleration of Autolysis During Fish Sauce Fermentation by Adding Acid and Reducing the Salt Content." *J. Sci. Food Agric.* 35:1363–1368.

GLICKSMAN, M. 1969. *Gum Technology in the Food Industry. Food Science and Technology—A Series of Monographs.* New York: Academic Press.

HAARD, N. F. 1992. "Biochemistry and Chemistry of Color and Color Changes in Seafoods." In Pp. 305–360. *Advances in Seafood Biochemistry: Composition and Quality,* edited by G. J. Flick, Jr. and R. E. Martin. Lancaster, PA: Technomic Publishing Co.

HEW, C. L., and FLETCHER, G. L. 1985. "Biochemical Adaptation to the Freezing Environment-Structure, Biosynthesis and Regulation of Fish Antifreeze Polypeptide." In Pp. 553–563. *Circulation, Respiration and Metabolism,* edited by R. Gilles. Berlin: Springer-Verlag.

HEW, C. L., and DAVIS, P. L. 1987. "Structure and Function of Fish Antifreeze Polypeptides. In Pp. 141–161. *Proteins: Structure and Function,* edited by J. J. L'Italian. New York: Plenum Press.

HOLAHAN, J. F. 1965. "Manufacture of Color Picture Tubes." *Electronics World* 74(6):30–32, 56.

KARRICK, N. L. 1976. "Fish Meat Quality." Pp. 245–253. In *Industrial Fishery Technology,* edited by M. E. Stansby. Huntington, NY: Robert E. Krieger Publishing Co.

KARUBE, J., and SODE, K. 1988. "Enzyme and Microbial Sensor." In Pp. 115–130. *Analytical Uses of Immobilized Biological Compounds for Detection, Medical and Industrial Uses,* edited by G. G. Guilbault and M. Mascini. Boston: Reidel Publishing Co.

KOMATA, Y. 1990. "Umami Taste of Seafoods." *Food Rev. Internat.* 6:457–487.

LEE, C. F. 1976. "Processing Fish Meal and Oil." Pp. 209–226. In *Industrial Fishery Technology,* edited by M. E. Stansby. Huntington, NY: Robert E. Krieger Publishing Co.

LEE, C. M. 1984. "Surimi Process Technology." *Food Technol.* 38(11):69–80.

LEE, Y. B., ELLIOT, J. G., RICHANSRUD, D. A., and HAGBERG, E. C. 1978. "Predicting Protein Efficiency Ratio by the Chemical Determination of Connective Tissue Content in Meat. *J. Food Sci.* 43:1359–1362.

LONG, A. M., and HAARD, N. F. 1988. "The Effect of Carotenoid–Protein Association on Pigmentation and Growth Rates of Rainbow Trout (*Salmo gairdner*)." *Bull. Aquaculture Assoc. Canada* 88(4):98–100.

MARTINEZ, A., OLSEN, R. L., and SERRA, J. L. 1988. "Purification and Characterization of Two Trypsin-like Enzymes from the Digestive Tract of an Anchovy *Engraulis encrasicholus*." *Comp. Biochem. Physiol.* 91B:677–684.

NORLAND, R. E. 1990. "Fish Gelatin." Pp. 325–333. In *Advances in Fisheries Technology and Biotechnology for Increased Profitability,* edited by M. N. Voigt and J. R. Botta. Lancaster, PA: Technomic Publishing Co.

OCKERMAN, H. W., and HANSEN, C. L. 1988. "Seafood By-products." Pp. 279–308. In *Animal By-Product Processing.* Weinheim, Germany: VCH Publishers.

OLSEN, R. L., JOHANSEN, A., and MYRNES, B. 1990. "Recovery of Enzymes from Shrimp Waste." *Process Biochem.* 25:67–68.

OLSEN, R. L., ØVERBØ, K., and MYRNES, B. 1991. "Alkaline Phosphatase from the Hepatopancreas of Shrimp (*Pandalus borealis*): A Dimeric Enzyme with Catalytically Active Subunits." *Comp. Biochem. Physiol.* 99B:755–761.

OSBORNE, R., VOIGT, M. N., and HALL, D. E. 1990. "Utilization of Lumpfish (*Cyclopterus lumpus*) Carcasses for the Production of Gelatin." Pp. 143–150. In *Advances in Fisheries Technology and Biotechnology for Increased Profitability*, edited by M. N. Voigt and J. R. Botta. Lancaster, PA: Technomic Publishing Co.

PIGOTT, G. M. 1986. "Surimi: The High Tech Raw Material for Fish Flesh." *Food Rev. Internatl.* 2:213–246.

RAA, J. 1990. "Biotechnology in Aquaculture and the Fish Processing Industry: A Success Story in Norway." Pp. 509–524. In *Advances in Fisheries Technology and Biotechnology for Increased Profitability*, edited by M. N. Voigt and J. R. Botta. Lancaster, PA: Technomic Publishing Co.

RAA, J., and GILDBERG, A. 1976. "Autolysis and Proteolytic Activity of Cod Viscera." *J. Food Technol.* 11:619–628.

RAA, J., GILDBERG, A. 1982. "Fish Silage: A Review." *Crit. Rev. Food Sci. Nutr.* 16:383–408.

RUBINSKY, B., ARAV, A., and FLETCHER, G. L. 1991. "Hypothermic Protection— A Fundamental Property of Antifreeze Proteins." *Biochem. Biophys. Res. Comm.* 180:566–571.

SHAHIDI, F., and SYNOWIECKI, J. 1991. "Isolation and Characterization of Nutrients and Value-Added Products from Snow Crab (*Chinoectes opilio*) and Shrimp (*Pandalus borealis*) Processing Discards." *J. Agric. Food Chem.* 39:1527–1532.

SHAHIDI, F., and SYNOWIECKI, J. 1992. "Nutrient Composition of Mechanically-Separated and Surimi-like Seal Meat." *Food Chem.* 47:41–46.

SHAHIDI, F., RUBIN, L. J., DIOSADY, L. L., and WOOD, D. F. 1987. "Preparation of the Cooked Cured-Meat Pigment, Dinitrosyl Ferrohemochrome, from Hemin and Nitric Oxide. *J. Food Sci.* 50:271–272.

SHAHIDI, F., NACZK, M., PEGG, R. B., and SYNOWIECKI, J. 1991. "Chemical Composition and Nutritional Value of Processing Discards of Cod (*Gadus morhua*)." *Food Chem.* 42:145–151.

SHAHIDI, F., HAN, X. Q., SYNOWIECKI, J., and BALEJKO, J. A. 1992a. "Enzymatic Modification of Marine Proteins: 1. Male and Spent Capelin. In *Book of Abstracts. Atlantic Fisheries and Technology Conference.* 23–26 August. Percé, Quebec, Canada.

SHAHIDI, F., SYNOWIECKI, J., and BALEJKO, J. A. 1992b. "Utilization of Seal Meat By-products." Pp. 1121–1124. In *Proceedings of the 37th International Congress of Meat Science and Technology.* August 23–28, Clermont-Ferrand, France.

SHAHIDI, F., SYNOWIECKI, J., DUNAJSKI, E., and CHONG, X. 1992c. "Non-protein Nitrogen Compounds in Harp Seal (*Phoca groenlandica*) Meat. *Food Chem.* 46:407–413.

SHAHIDI, F., SYNOWIECKI, J., and PENNEY, R. W. 1993. "Pigmentation of Arctic Char (*Salvelinus alpinus*) by Dietary Carotenoids." *J. Aquatic Food Prod. Technol.* 2:99–115.

SHAMSUZZAMAN, K., and HAARD, N. F. 1983. "Evaluation of Harp Seal Gastric Protease as a Rennet Substitute for Cheddar Cheese. *J. Food Sci.* 48:179–182.

SIMPSON, B. K., and HAARD, N. F. 1987. "Cold-adapted Enzymes from Fish." Pp. 495–527. In *Food Biotechnology*, edited by D. Knorr, New York: Marcel Dekker.

TERASAKI, M., KAJIKAWA, M., FUJITA, E., and ISHII, K. 1965. "Studies on the Flavor of Meats. Part 1. Formation and Degradation of Inosinic Acid in Meats." *Agric. Biol. Chem.* 29:208–212.

WARD, A. G., and COURTS, A. 1977. *The Science and Technology of Gelatin.* New York: Academic Press.

13

Biotechnological Applications of Seafood Proteins and Other Nitrogenous Compounds

Norman F. Haard, Benjamin K. Simpson, and Zdzisław E. Sikorski

INTRODUCTION

Marine biotechnology is a rapidly growing field that has taken on new meaning as a result of recent advances in natural product chemistry, comparative biochemistry, molecular biology, and gene technology. In terms of protein modification, traditional marine biotechnology can be likened to a "sledge hammer approach." Fermentation processes, like fish sauce, fish silage, "biological" fish protein concentrate, matjes herring, and fermented squid, extensively alter protein and other constituents by endogenous microbial action, by autolytic enzymes, or by ill-defined enzyme additives. These "artlike" processes are time-honored ways to preserve and/or transform underutilized fish and fishery by-products into useful food and feed products. Modern marine biotechnology includes the transformation and/or recovery of "value-added" specialty products from marine organisms. For example, fish protein may be treated with particular proteinases to produce specific emulsifiers, surfactants, flavorings, and pharmaceuticals. Natural products from marine organisms may possess unique properties. Examples of products that may be recovered from marine organisms include enzymes, vaccines, antibiotics, amino acids, antifreeze proteins, and hormones.

The oceans occupy 70% of our planet's surface area and contain a diverse and abundant taxonomic array of organisms that include microorganisms, plants, and endothermic and ectothermic animals. The bony fishes alone are composed of 50 taxonomic orders with approximately 40,000 species. The unique characteristics of marine organisms and their cellular components reflect their environmental habitats. The ocean environment is aqueous, buoyant, saline, cold (mostly 5°C or less), and highly variable in illumination, pH, water movement, and pressure. It is, therefore, not surprising that those marine organisms collectively possess a plethora of distinctive molecules with peerless properties in the terrestrial environment.

BIOCHEMICALS FROM MARINE ORGANISMS

Enzymes

Introduction. The gross sale of industrial enzymes was estimated to be $600 million per annum in 1985. The use of enzymes in food processing and other industries is reviewed by Godfrey and Reichert (1983). Approximately 90% of all industrial enzymes are hydrolases used for depolymerization of natural substances. The current distribution of industrial enzymes is summarized in Table 13–1. Two basic methods for manufacturing industrial enzymes are (1) microbial fermentation and (2) extraction from plant or animal biomass. Genetic engineering techniques may be used to produce plant or animal enzymes using bacteria or yeast expression systems. The main obstacles to the use of marine by-products as sources of industrial enzymes include their limited availability due to harvest fluctuation, variability in levels or activities of enzymes due to intraspecific factors (Chapter 3), and the instability or perishable nature of the raw material.

Choice of Enzyme. The choice of enzyme for a given application depends on a variety of factors related to the chemical nature of the material being treated and the intended process. The main factors considered by industrial enzymologists are catalytic specificity, pH optimum, pH stability, temperature optimum, thermal stability, and the enzyme's response to inactivating or inhibiting substances such as salt, acid, alkali, tannins, detergents, and specific natural enzyme inhibitors. Homologous enzymes from different sources, which catalyze the same basic reaction, may differ with respect to several of these factors. For example, uricase

Table 13–1   Industrial uses of enzymes.

Type	Enzyme	Industrial Use (%)
Proteases	All groups	59
	Trypsins	3
	Rennets	10
	Acid proteases	3
	Neutral	12
	Alkaline	31
Carbohydrases	Pectinases	3
	Isomerases	6
	Lactase	1
	α-Amylases	5
	β-Amylases	13
Lipases	Microbial	3
Others	Analytical, pharmaceutical, developmental	10

Source: Adapted from Godfrey and Reichert (1983).

catalyzes oxidation of uric acid to allantoin. The use of this enzyme in clinical measurement of serum uric acid is limited by the instability of current industrial sources at neutral pH. Because trout liver uricase is stable at pH 7, it may be more useful than mammalian sources of uricase for this particular application (Kinsella et al., 1985). Applications involving proteases are often limited by the inability to produce peptides of definite size (Chobert et al., 1992). For example, proteases that catalyze a high degree of hydrolysis can lead to bitter peptides in cheese and protein hydrolysates, loss of "head" stability in chill-proofed beer, over-tenderization of meat, and stickiness and poor texture in bread from dough treated to reduce mixing time.

Cold Adapted Enzymes from Fish.   Poikilotherms living in a cold-water habitat exhibit enzyme adaptability in a variety of ways: (1) increased enzyme concentration in the cell, (2) changes in the type of enzyme present in the system, and (3) adaptability of preexisting enzymes. Cold acclimation may be manifest in alterations in isoenzyme expression, substrate binding, substrate turnover rate, thermal stability, physiological efficiency, and thermodynamic properties [reviewed by Simpson and Haard (1987a); Haard 1992]. Cold-adapted enzymes have several advantages as industrial enzymes, namely, processes may be run at a reduced reaction temperature, thus minimizing bacterial activity and competing reactions, and the enzyme may be readily inactivated at the completion of the process. At low temperatures, lysozyme from clam shell is several hundred times more active than lysozyme from warm-blooded animals (Raa, 1990).

Digestive Enzymes. Enzymes with unique properties for industrial usage can be recovered as by-products from fish processing wastes (Haard et al., 1982; Haard et al., 1983; Simpson and Haard, 1987a; Stefansson and Steingrimsdottir, 1990; Almas, 1990; Raa, 1990; Haard, 1992; Simpson et al., 1991). A number of laboratories are now developing commercial processes for isolating enzymes from fish processing offal. The Icelandic Fisheries Laboratory has developed a process to recover trypsinlike enzymes from cod viscera (Stefansson and Steingrimsdottir, 1990). The viscera are extracted at neutral or slightly alkaline pH. The extract is sieved, and proteins are precipitated and separated from low-molecular-weight impurities by ultrafiltration. The crude enzymes can be further fractionated to yield trypsin, chymotrypsin, and elastase. Enzyme isolates are stabilized by freeze-drying or spray-drying. Marine Biochemicals in Tromso, Norway has developed industrial processes for the recovery of pepsin, trypsin, chymotrypsin, alkaline phosphatase, and hyaluronidase from fish viscera (Almas, 1990). Fish viscera in frozen blocks is purchased from the fish processing industry for extraction of digestive enzymes (Raa, 1990). Methods for isolating marine enzymes from marine resources are also being developed in Japan (Seto, 1990), Denmark (Kristjansson, 1990; Borresen, 1992), and Great Britain (Reece, 1988). A crude enzyme preparation obtained by dehydration and defatting of cod and mackerel pyloric ceca by a methanol : ether (9 : 1) mixture was successfully applied in bating fish skins in the tanning process (Trzesinski, 1960). Pyloric ceca from salmon cannery waste are defatted and dried with acetone : ether (9 : 1 v/v) to make a powder used in leather bating (Ockerman and Hansen, 1988). Alkaline phosphatase is recovered from the thaw drip of frozen shrimp by ultrafiltration and is used in diagnostic kits (Olson et al., 1990).

Collagenolytic Enzymes. Collagenolytic enzymes have been extracted from the hepatopancrease of marine invertebrates such as crab, prawn, and lobster (Eisen and Jeffrey, 1969; Eisen et al., 1973; Grant et al., 1983; Baranowski et al., 1984; Nip et al., 1985; Chen, 1992). These enzymes are thought to produce mushiness in the flesh of these animals during postharvest handling and storage. However, industrial applications of specific collagenolytic enzymes may be sought in the food (meat tenderizing) and leather (bating and dehairing) industries. Traditional proteinases used as meat tenderizers are mainly from plant origin, namely papain, ficin, and bromelain (Bernholdt, 1975; Dransfield and Etherington, 1981). But these proteinases currently used for tenderization attack both connective tissue and myofibrillar proteins, thus they can bring overtenderization, leading to mushiness and undesirable texture softening. The ideal enzyme for degrading insoluble collagen in a meat system must not only demonstrate specificity for collagen, but must also function at (1) the relatively low pH of meat, (2) the low temperature at which meat is held prior to cooking, or (3) at the high temperatures reached during cooking. Collagenases have been found to have the desired properties, and especially collagenases from marine sources have been shown to offer high-quality meat and meat products, and their use is encouraged for commercial application (Cronlund and Woychik, 1987).

Polyphenol Oxidases. Polyphenol oxidases (PPO) in crustacean species are responsible for the postharvest dark discoloration's in these animals (Opoku-Gyamfua et al., 1992; McEvily et al., 1991; Ferrer et al., 1989; Simpson et al., 1987, 1988; Bailey et al., 1960). Their counterparts in plants are essential in the reactions involved in fermentation of black tea. The properties of the enzymes derived from marine organisms (i.e., higher activity at low reaction temperature, and low thermal stability) appear to be better suited for tea fermentation than plant PPOs. For example, the fermentation temperature significantly affects the total pigmentation of black tea; and the optimal temperature for the tea fermentation process is 30°C (Robertson, 1991). The temperature optimum of PPO's from marine organisms are in the range 30–40°C and are generally lower than plant PPOs (Opoku-Gyamfua et al., 1992). The disadvantages of using plant PPOs for tea fermentation include the following: (1) at temperatures above 40°C, high enzyme activity combined with low oxygen solubility severely inhibit the fermentation process and (2) the enzyme tends to be present in tea leaves at varying levels, thus there is the need to closely monitor and/or vary the conditions of the fermentation process from batch to batch to ensure tea products with consistent properties. Modern methods of tea processing aim at more efficient fermentation, thus necessitating manipulation of processing conditions to provide black teas of different compositions for particular markets. One such manipulation could take advantage of recent advances in genetic engineering to exploit the unique properties of PPOs from marine organisms for tea fermentation.

Gelatin

Fish gelatin is manufactured from the skins of cold-water fish such as cod, haddock, pollack, hake, and cusk (Norland, 1990). The raw skins are washed in running water for 3–4 h, soaked in dilute alkali for 6–8 h, again washed in running water for 3–4 h, washed in dilute sulfurous acid, and again washed in running water. The gelatin is then extracted with water at 70–80°C (Ockerman and Hansen, 1988). The uniqueness of fish gelatin from cold-water fish lies in the lower content of amino acids, proline, and hydroxyproline. These fish gelatins form gels at 8°C, whereas land-animal gelatins form gels at above 30°C. Although fish gelatin does not form particularly strong gels, it is well suited for certain industrial applications, for example, microencapsulation of dye precursors for carbonless paper, light-sensitive coatings on photographic film and other surfaces, and high-strength, low-set-time glue. Isinglass, made from the swim bladder of fish, is a high-grade fish gelatin used as a clarifying agent for cider, wine, beer, and vinegar. Skins from fish can also be converted into leather by a tanning procedure similar to the one used for land-animal hides (Trzesiński, 1955). Fish skins produce leather with excellent properties but do not occupy a large part of the leather market because they are not economically competitive with synthetics and land-animal leathers (Ockerman and Hansen, 1988).

Albumin

Albumin isolated from low-fat fish, mollusks, and crustaceans can be used as an egg albumin replacer [reviewed by Ockerman and Hansen (1988)]. Minced flesh is cooked in 0.5% acetic acid for 1 h at 80–90°C to extract most of the connective tissue. The residue is washed to remove acid and digested protein, pressed, ground, and extracted for 6–8 h with ether, alcohol, or trichloroethylene to remove fat. The dried residue is digested with NaOH, neutralized with lactic acid, and spray-dried. The product, containing primarily polypeptides, is used in place of egg albumin as a whipping, suspending, or stabilizing agent.

Pharmaceuticals

Insulin, used in the treatment of diabetes, is generally extracted from the pancreas of cattle. Insulin has been isolated from fish and is generally more stable than that from bovine pancreas after extraction. The Langerhans islets of various fishes, that is, cod, halibut, pollack, bonito, and tuna, are used as a source of insulin. Islets are crushed in picric acid solution, and extraction with alcoholic-HCl yields water-soluble insulin hydrochloride. The tissue may be preserved in 95% ethanol acidified with 0.3% HCl or frozen fresh prior to extraction (Ockerman and Hansen, 1988; Wheaton and Lawson, 1985).

Antimicrobials, Antitumor Compounds, Antioxidants. A number of other protein and nonprotein natural products may be recovered from marine organisms. Extracts from mussels and clams contain antibacterial substances that specifically inhibit certain *Vibrio* sp. These substances are being developed to cure bacterial diseases in farmed salmon (Raa, 1990). Seaweed's contain antiviral, antifungal, antihelminthic, and antibiotic substances (Chapman and Chapman, 1980).

Natural hydroquinones with physiological activity similar to α-tocopherol have been identified in marine tunicates (Cotelle et al., 1991). The substances appear to react directly with superoxide anion.

Anti-tumor compounds (alkaloids) have been identified as natural marine products from sponges (Burres et al., 1989; 1991; Burres and Clements, 1989). Three small peptides, didemins A, B, and C, isolated from the marine tunicate *Trididemnum* sp. have been shown to possess strong antitumor activity (Rinehart et al., 1988). The topic of marine natural products and medicine have been reviewed (Scheuer, 1989; Faulkner, 1990).

Other Substances

DNA and Nucleotides. Nucleotides and nucleic acids can be produced from testes, and protamines from salmon milt (Ockerman and Hansen, 1988). DNA has been produced from fish milt for uses in the cosmetics industry. Protamines, combined with 5-iododeoxyuridine, are used for treatment of terminal cancer patients. Nucleic acids isolated from fish milt can be hydrolyzed to taste active nucleotides for use as food additives (Almas, 1990) and used in the biochemical industry (Gildberg and Almas, 1986).

Deodorizing Substances. Deodorizing substances have been isolated from edible seaweed in Japan (Seto, 1990). For example, phlorotannin isolated from *E. bicyclis* is a more effective deodorant than chlorophyll (Nakajima and Sibuya, 1988).

Lectins. A group of proteins called lectins are being isolated from marine invertebrates for use in medical diagnostics and separation of cells (Raa, 1990). Lectins have the ability to specifically bind to carbohydrates, glycoproteins, and glycolipids on cell surfaces.

Chitin and Chitosan. Chitin is conjugated with protein in the exoskeletons of marine invertebrates, insects, and arachnids (Austin et al., 1981), and in the cell walls of most fungi and algae. It is the second most abundant organic compound in nature after cellulose (Ruiz-Herrera, 1978), and it is a homopolymer of $\beta(1-4)$-linked N-acetylglucosamine residues. The molecular structure of chitin is similar to that of cellulose, except that the –OH groups bonding the second carbon position of cellulose are substituted into acetamide groups in chitin (Tsugita, 1990). Shellfish comprising shrimp, crab, lobster, prawn, and crayfish contain 14–35% chitin on a dry weight basis. Crude chitin, containing 6.3% nitrogen, makes on the average 40% of the deproteinized, dry shell waste of Antarctic krill (Naczk et al., 1981). These source materials for chitin are becoming a growing waste disposal problem for the food and biotechnology industries (Revah-Moiseev and Carroad, 1981). It is estimated that more than 100 billion tons/annum of chitin are produced by these organisms. Unfortunately, chitin has been put to very little use by man. The amount of annually accessible chitin has been estimated at 150,000 tons (Allan et al., 1978), and total sales of chitin/chitosan are expected to reach almost 2 billion U.S. dollars during the next 10 years (Knorr, 1991). Current procedures for the preparation of chitin from shellfish waste consists of two steps: protein separation or deproteinization with dilute alkali or proteolytic enzymes, followed by $CaCO_3$ separation or demineralization with dilute acid or chelating agents (Sikorski and Naczk, 1982b; Tsugita, 1990).

Chitosan is a copolymer of glucosamine and N-acetylglucosamine units. It is derived from chitin by deacylation, that is, the acetamide groups in chitin are substituted by amino groups in chitosan. Deacylation may be achieved by chemical means (i.e., with hot concentrated NaOH solution) or by enzymatic means using chitin deacylase produced by microorganisms like *Mucor rouxii, Phycomyces blakesleeanus,* and *Aspergillus niger.* Chitosan is a positively charged polyelectrolyte which makes it a more useful and valuable product than chitin. The primary difference between chitin and chitosan is solubility. Chitosan, unlike chitin, is soluble in dilute acid solutions; however, both are insoluble in alkaline solutions and organic solvents.

The chemical procedures for producing chitin and chitosan result in commercial products of inconsistent physicochemical characteristics because of the rather nonspecific nature of the action of these chemicals (Knorr and Klein, 1986). The enzymatic approach recovers products with better and more consistent properties; however, the yields are rather low.

The advantages of using chitinous polymers as food and other processing aids include the following: They are natural, nontoxic components of foods, are biodegradable and have been approved for use as food ingredient by the FDA in the United States; and are used as clarifying agents for wines and drinking water in the United Kingdom and other European countries. Chitinous polymers also have antimicrobial properties and other health-related benefits, such as their capacity to tie up bile salts, as well as their ability to lower blood serum cholesterol levels. The manifold applications of chitin and chitosan polymers include their use in (1) the food industry for biomass recovery as in cheese processing; sludge treatment in brewing; as packaging films or edible food wraps; in flavor production; and as a food thickener and/or stabilizer; (2) in paper technology as a papermaking additive for paper surface strength improvement; (3) in water pollution for the removal of color; in wastewater sludge treatment; and for removal of heavy metals and environmental contaminants like PCBs; (4) in plant protection as defense against pathogens; inhibition of fungal growth; and the suppression of plant parasites; (5) in animal nutrition as feed supplement; dietary fiber; digestive aid; and for the promotion of growth of *bifidobacteria;* (6) in the chemical industry as oil refinery wastewater decontaminant; as molecular sieves; as flame retardent; in reverse osmosis; as an adhesive; and for enzyme or cell immobilization; (7) in cosmetics in dry shampoos; hair conditioners and hair setting lotions; (8) in pharmaceuticals as a diluent for drug and other tablet manufacture; and (9) in biomedicals as bile salt sequesterant; surgical sutures; vascular grafts; anticoagulant; aggregation of cancer cells; and as artificial kidney membrane. The potential for the use of chitin as an enzyme support has been shown by Synowiecki et al. (1981; 1982) and for the application of chitosan powder in a biosensor for monitoring lipid oxidation in food systems was recently demonstrated by Weist and Karel (1992).

PROCESSING SEAFOOD PROTEINS WITH ENZYMES

Traditional Processes

Introduction.   Until recently, the commercial use of enzymes in the seafood industry has been limited to a few applications. These include reduction of stickwater viscosity in the fish meal industry, preparation of fish protein hydrolysates for feed or animal use, and fermentation processes depending on endogenous and contaminating microbial proteases. The subject has been reviewed recently (Stefansson and Steingrimsdottir, 1990; Haard, 1992; Borresen, 1992).

Stickwater Treatment.   In the fish meal and canning industries, dissolved protein that is concentrated by evaporation to make a concentrate called "fish solubles" can become viscous and prevent efficient evaporation. The addition of neutral or alkaline bacterial proteases, when the solids content is above 20%, results in fuel savings and product improvement (Jacobson and Lykke-Rasmussen, 1984).

Biological Fish Protein Concentrate.   Fish protein concentrate for human consumption produced by solvent extraction methods has had limited acceptance in the market. The functional properties of the concentrates can be improved by limited proteolysis, induced by enzyme treatment or by an alkaline process (Sikorski and Naczk, 1981; 1982a). Enzymatic hydrolysis has been extensively used in producing fish protein preparations from the small pelagic fish Caspain sprats (Chernogortsev, 1973). FPC produced by an enzyme-assisted process has been called "biological fish protein concentrate." The process normally involves limited hydrolysis of inexpensive fish or processing wastes with proteases such as ficin or papain, inactivation of the enzyme, filtration of the hydrolysate, concentration, and spray-drying. A problem with FPC is the formation of bitter tasting hydrophobic peptides (Mohr, 1980). The problem can be avoided by using proteases with appropriate specificity (Simpson and Haard, 1987a; Rebeca, Pena-Vera, and Diaz-Castaneda, 1991; Chobert et al., 1992).

Products Obtained by Protein Autolysis.   A number of traditional fishery products are prepared by fermentation's where protein hydrolysis is the predominant reaction. These products include fish sauce, cured fish products, and fish silage.

Fish sauce is prepared by allowing fish to slowly digest in the presence of 20–30% salt (Orejano and Liston, 1981). After about 1 year, a clear liquid containing free amino acids and small peptides is recovered by filtration. Large peptides (> 10 kD) also contribute to the flavor of the traditional product (Raksakulthai and Haard, 1992). The addition of industrial enzymes can accelerate the process; however, the flavor characteristic of the finished product is normally inferior to the traditional product. Supplementation of capelin fermentation with enzymes from squid hepatopancreas accelerates the process and yields a product with superior sensory properties (Raksakulthai et al., 1986).

Fish silage is a liquid protein hydrolysate produced from fish processing waste and small pelagic fish under acidic conditions. The autolysis depends on endogenous proteinases from both the viscera and the flesh (Haard et al., 1985). The product is used a protein supplement for animal feed (Raa and Gildberg, 1982; Gildberg and Almas, 1986; Ockerman and Hansen, 1988).

Fermented products in which less-extensive protein hydrolysis occurs include cured herring (Stefansson and Steingrimsdottir, 1990) Matjes herring (Simpson and Haard, 1984) and squid (Lee et al., 1982). Intact fish or meat sections are brined and stored for as long as 1 year with the traditional process. During ripening, endogenous, digestive enzymes diffuse into the muscle and proteolysis changes the taste characteristics without greatly altering texture. The role of nitrogenous compounds in fermented and salted fishery products has been extensively treated in Chapter 9. Supplementing the fermentation with herring digestive enzyme (Eriksson, 1975), bovine trypsin or Greenland cod trypsin (Simpson and Haard, 1984) accelerates the ripening process. Lipase action also appears to contribute to the ripening of Matjes herring (Ritskes, 1971). According to Borresen (1992), commercial enzyme mixtures are now used by the cured herring industry. Squid hepatopancreas extract containing cathepsin C accelerates ripening of fermented squid and increases the sweet taste of the product (Lee et al., 1982).

Recent Developments

Recent enzyme applications for the seafood industry have been reviewed (Stefansson and Steingrimsdottir, 1990; Simpson et al., 1991; Borreson, 1992; Haard, 1992).

Skin Removal.   A number of inventions use proteases to remove skin from fish. Tuna skin removal is facilitated with a complex mixture of proteases and carbohydrases (Fehmerling, 1973). Herring skin removal is accomplished with cod pepsin (Joakimsson, 1984). Squid is skinned with papain (Raa, 1990; Hempl, 1983) and skate skin is removed with an unspecified mixture of protease(s) and carbohydrase(s) (Wray, 1988; Stefansson and Steingrimsdottir, 1990). Research has also shown it possible to loosen shrimp shells from the meat using enzymes (Raa, 1990).

Scale Removal.   Mechanical methods to remove scales may result in tearing of skin and reduced fillet yield. In Iceland, fish scale removal using enzymes has recently been accomplished (Stefansson and Steingrimsdottir, 1990). The unspecified enzymatic removal of scales from ocean perch and haddock is done at 0°C.

Membrane Removal.   Proteases, including fish pepsin, can be used to hydrolyze the supportive tissue that envelops roe sacks (Sugihara et al., 1973; Gildberg and Almas, 1986; Raa, 1990). Pepsins appear to split key linkages that adhere the egg cells to the roe sack without affecting the eggs. The process has been used commercially in Canada and Scandinavia (Wray, 1988) with a twofold increase in roe yield (Anonymous, 1990).

Other applications of proteases to remove membranes include treatment of cod liver prior to canning (Stefansson and Steingrimsdottir, 1990), removal of the black membrane surrounding the swim bladder

(Steingrimsdottir and Stefansson, 1988), and removal of intestines from scallops (Stefansson et al., 1987; Raa, 1990).

Flavor Isolation. Proteolytic enzymes have also been used to recover flavor compounds from shrimp (Chen and Li, 1988; Pan, 1990). Active enzymes in the cephalothorax of grass shrimp that contribute to flavor extraction include carboxpeptidases A and B, trypsin, chmotrypsin, and collagenase (Chen and Li, 1988). Shrimp flavorant obtained by enzymatic processes can be used in surimi-based and cereal-based extrusion products. Seafood flavors are produced commercially by an enzymatic process in France (In, 1990).

Pigment Isolation. Trypsin has been used to recovery carotenoprotein from shell waste of shrimp (Simpson and Haard, 1985; Cano-Lopez et al., 1987) and crab (Manu-Tawiah and Haard, 1987). About 80% of the protein and 90% of the astaxanthin can be recovered from shrimp waste as a stable lipoprotein complex. When used in fish feed, carotenoprotein pigment is more efficiently deposited in tissues of cultured trout than free astaxanthin (Long and Haard, 1988).

Tenderization of Squid. An enzyme preparation from squid hepatopancreas has been used to tenderize squid (Kołodziejska et al., 1992). In squid mantle meat, treated with the extract 24 h at 4°C, the sarcoplasmic and myofibrillar proteins were hydrolyzed to low-molecular-weight fractions. Soaking isolated collagen fibers 24 h in the liver extract resulted in almost complete solubilization of the fibers in buffered SDS/urea solution. The solubilized product contained only low-molecular-weight components. The cooked squid mantle treated with the enzyme preparation was considerably more tender than that of untreated samples.

Enzyme Inactivation. Trypsinlike enzymes from stomachless marine organisms have been shown to be effective in inactivation of enzymes (Pleiderer et al., 1967; Simpson and Haard, 1987b). A suggested application of trypsins, which readily catalyze hydrolysis of native enzymes, is prevention of "black spot" by inactivation of shrimp polyphenoloxidase (Simpson et al., 1991).

Production of Peptones. Peptones are partially hydrolyzed proteins, which are soluble in water, not heat coagulable, and used in bacterial culture media. Investigations using fishery resources for peptone production are summarized by Iyer et al. (1978), Strom and Raa (1991), Vecht-Lifskitz et al. (1990), and Ockerman and Hansen (1988).

Soluble fish extract for microbiological media was prepared from menhaden "fish solubles" by defatting with hexane, centrifuging, filtering, and freeze-drying (Green et al., 1977). The same authors prepared soluble hake peptone by autolysis of hake at 50–55°C, clarification, and spray-drying. Both products supported comparable growth of various microorganisms, including *lactobacilli*, as commercial peptones.

Using papain-catalyzed hydrolysis of mackerel offal flesh, with cysteine and sodium EDTA, a 12% yield of peptone was prepared (Rao et al., 1978). The product compared favorably to commercial beef peptone in supporting growth of *Lactobacillus arabinosus* and *Streptococcus faecalis*. The same authors prepared peptone with shrimp heads using papain hydrolysis, with BHA, EDTA, chloroform, and toluene, for 4 h at 45–50°C, pH 6.5–7.0 (Rao et al., 1980). The filtered hydrolyzate was vacuum concentrated to 20–25% solids and vacuum dried to 5.7% moisture. The isolate supported similar growth of *Streptococcus faecalis* and *Leuconostoc mesenteroides* as oxoid peptone. Papain has also been used to prepare peptone from whole, minced *Nemipterus japonicus* (Iyer et al., 1978).

Bovine and porcine pancrease tissue have also been used to prepare peptone from the flesh of cod, blue whiting, and grenadier (Dunajski et al., 1987). The optimum proportion of pancreas to fish meat was 1 : 5 w/w for a 3-h hydrolysis time. The fish peptones supported similar growth of *Escherichia coli*, *Shigella sonnei*, and *Staphylococcus aureus* as Difco peptones. More recent studies have shown that pancreatic enzyme (1 : 5 w/w) catalyzed hydrolysis of parasitized cod fillets for 6 h at 40°C, pH 7.0, gave a peptone that supported growth of *E. coli* and *S. aureus* better than Bacto-Tryptone, Bacto-Tryptose, Bacto-Peptone, and Proteose-Peptone (Skorupa and Sikorski, 1992).

GENETIC ENGINEERING

Introduction

Application of recombinant DNA technology and genetic engineering to aquaculture species have developed rapidly since 1985. These new biotechnology techniques will undoubtedly be important to the fishery and aquaculture industries in the future. Problems with fish production (e.g., growth, disease, and nutrition) and with seafood quality (e.g., color, flavor, texture, safety, nutrition, stability, and processing suitability) can be addressed by genetic engineering in combination with traditional genetic improvement programs. We should emphasize, however, that the ability of the genetic engineer to improve fish production and quality characteristics is dependent on knowledge of basic biology and food biochemistry. For example, for the genetic engineer to develop a transgenic fish with less postharvest flesh softening, there is need for basic biochemical information on the mechanism of texture change. There are reservations about the future of transgenic foods because of concerns with government regulation, food safety, and environmental risk. In the United States, the first genetically engineered food was recently approved by the Food and Drug Administration (Gershon, 1992). Indeed, FDA expects that many gene-altered foods will not require premarket approval.

Transgenic Fish

Several reviews have been published on the subject of transgenic fish (Mclean et al., 1987; Tave, 1988; Hew, 1989; Chen and Powers, 1990; Hallerman et al., 1990; Houdebine and Chourrout, 1991). Within the past 7 years, microinjection gene-transfer techniques have become available to the fish breeder. In most fish species, external fertilization is a natural process. Thus, injections of embryos with cDNA does not require complex manipulations needed in mammalian systems, such as in vitro culture and transfer of embryos into foster mothers. The survival rate of injected fish embryos is high (35–80%) compared to mammalian embryos. Although cytoplasmic microinjection is a successful method to produce transgenic fish, it is tedious and unsuitable for mass gene transfer. Therefore, other DNA delivery systems such as electroporation, viral infection, liposomes, and particle gun microprojectiles are being investigated.

Transfer of foreign genes into fish embryos by microinjection usually results in random integration of gene sequences into the host genome. Transgenic fish produced by this method are "mosaics" that do not possess the foreign DNA in every cell. Methods for targeting foreign genes to specific areas of the genome are needed because integration of foreign genes can interrupt transcription by the host genes, resulting in embryonic abnormality. However, many transgenic fish still transmit the stable integrated DNA into their progeny. A proportion of the F_1 population has the genes in all of their somatic cells and the parent germ cells and usually transmit the foreign DNA to their progeny in expected Mendelian ratios. Transgenic medaka-containing growth hormone and mMT promoter by electroporation transmitted 100% of the transgene to F_1 offspring and 88% to F_2 offspring (Inoue et al., 1990).

The formation of sufficient quantities of the foreign gene's product in the transgenic organism is sometimes a limitation of this technique. Numerous promoters (constitutive and inducible) have been evaluated in transgenic fish. Regulatory gene sequences are also important because transgenic fish commonly produce the foreign gene product in tissues where expression is not normally found or desired.

A summary of gene transfers accomplished in fish at the time of this writing is shown in Table 13–2. Genes introduced into fish include human (hGH), bovine (bGH), trout (tGH), salmon (sGH) or rat growth hormone (rGH), chicken crystalline protein (cCRY), *E. coli* β-galactosidase (β-gal), hygromycin resistance (hyg), chloramphenicol transacetylase (CAT), and winter flounder antifreeze protein (AFP). Promoter genes used in constructs include mouse metallothionein (mMT), antifreeze protein (AFP), Rous sarcoma virus (RSV), simian virus 40 (SV40), and firefly luciferase (fLuc). Positive biological effects of foreign growth hormone gene constructs have been observed in transgenic fish. Loach,

Table 13–2 Transgenic fish produced by microinjection of gene constructs.

Fish	Structural Gene	Promoter Gene	Year
Common carp	rGH	RSV	1989
	hGH	mMT	1987
	tGH	RSV	1988
Chinese carp	hGH	mMT	1989
Catfish	rGH	RSV	1989
	tGH	RSV	1989
	hGH	mMT	1987
Goldfish	neo	SV40	1988
	hGH	mMT	1985
Loach	hGH	mMT	1986
Medaka	CRY	SV40	1986
	rGH	mMT	1989
	tGH	mMT	1989
Salmon	β-gal	mMT	1988
	hGH	mMT	1989
	AFP	AFP	1988
Tilapia	hGH	mMT	1988
	CAT	SV40	1988
Trout	hGH	mMT	1987
	rGH	mMT	1987
	hGH	SV40	1986
Wuchang fish	hGH	mMT	1987
Zebra fish	hyg	SV40	1988
	CAT	RSV	1990

Note: Refer to Hallerman et al. (1990) and Chen and Powers (1990) for specific references.

common carp, crucian carp, Atlantic salmon, channel catfish, medaka, and northern pike containing either human, bovine, or trout growth-hormone genes are reported to grow 10–80% faster than nontransgenic fish if proper promoters are utilized.

Although the early studies in this area have focused on incorporating growth hormone, future research is likely to explore incorporation of genes coding for other proteins important to the food quality and stability of fishery products. Isolation of genes for production of n-3 fatty acid synthesis, carotenoid pigments, antifreeze protein, flavor compounds, natural antioxidants, and so on, can lead to seafood products with utility as food.

REFERENCES

ALLAN, G. G., FOX, J. R., and KONG, N. 1978. "A Critical Evaluation of the Potential Sources of Chitin and Chitosan." In *Proceeding of the 1st International Conference on Chitin/Chitosan,* edited by R. A. A. Muzzarelli and E. R. Parisen. Cambridge, MA: M.I.T.

ALMAS, K. A. 1990. "Utilization of Marine Biomass for Production of Microbial Growth Media and Biochemicals." Pp. 361–372. In *Advances in Fisheries Technology and Biotechnology for Increased Profitability,* edited by M. N. Voigt and J. R. Botta. Lancaster, PA: Technomic Publishing Co.

ANONYMOUS. 1990. "Enzyme Breaks Up Trout Roe." *Fish Farming Internat.* 17:71.

AUSTIN, P. R., BRINE, C. J., CASTLE, J. E., and ZIKAKIS, J. P. 1981. "Chitin: New Facet of Research." *Science* 212:749–753.

BAILEY, M. E., FIEGER, E. A., and NOVAK, A. F. 1960. "Physico-chemical Properties of the Enzymes Involved in Shrimp Melanosis." *Food Res.* 25:557–564.

BARANOWSKI, E. S., NIP, W. K., and MOY, J. H. 1984. "Partial Characterization of a Crude Enzyme Extract from Freshwater Prawn, *Machrobrachium rosenbergii.*" *J. Food Sci.* 49:1494–1495.

BERNHOLDT, H. F. 1975. "Meat and Other Proteinaceous Foods." Pp. 473–492. In *Enzymes in Food Processing,* 2nd ed., edited by G. Reed. New York: Academic Press.

BORRESEN, T. 1992. "Biotechnology, By-products and Aquaculture." Pp. 278–287. In *Seafood Science and Technology,* edited by E. G. Bligh. Cambridge, MA: Blackwell Scientific Publications.

BURRES, N. S., BARBER, D. A., GUNASEKERA, S. P., SHEN, L. L., and CLEMENT, J. J. 1991. "Antitumor Activity and Biochemical Effects of Topsentin. *Biochem. Pharmacol.* 42:745–751.

BURRES, N. S., and CLEMENT, J. J. 1989. "Antitumor Activity and Mechanism of Action of the Novel Marine Natural Products Mycalamide-A and -B and Onnamide." *Cancer Res.* 49:2935–2940.

BURRES, N. S., SAZESH, S., GUNAWARDANA, G. P., and CLEMENT, J. J. 1989. "Antitumor Activity and Nucleic Acid Binding Properties of Dercitin, a New Acridine Alkaloid Isolated from a Marine *Dercitus* Species Sponge." *Cancer Res.* 49:5267–5274.

CANO-LOPEZ, A., SIMPSON, B. K., and HAARD, N. F. 1987. "Extraction of Carotenoprotein from Shrimp Process Waste with the Aid of Trypsin from Atlantic cod." *J. Food Sci.* 52:503–504 and 506.

CHAPMAN, V. J., and CHAPMAN, D. J. 1980. *Seaweeds and Their Uses.* 3rd ed. New York: Chapman and Hall.

CHEN, H. Y., and LI, C. F. 1988. Isolation, partial purification, and application of proteases from grass shrimp heads. *Food Sci. (China)* 15(3):230–243.

CHEN, T. T., and POWERS, D. A. 1990. "Transgenic fish." *Trends Biotechnol.* 8:209–215.

CHEN, Y. 1992. *Characterization of Semi-Purified Collagenase Fraction from Lobster Hepatopancreas.* M.Sci. thesis. McGill University, Montreal, Canada.

CHERNOGORTSEV, A. P. 1973. *Pererabotka Melkoi Ryby na Osnovie Fermentirovaniya Syrya.* Moscow: Pischchevaya Promyshlennost.

CHOBERT, J. M., SITHONY, M., and WHITAKER, J. R. 1992. Proteolytic degradation of food proteins. In *Advances in Seafood Biochemistry*, edited by G. J. Flick and R. E. Martin. Pp. 291–304, Lancaster, PA: Technomic Co., Inc.

COTELLE, N., MOREAU, S., BERNIER, J. L. CATTEAU, J. P., and HENICHART, J. P. 1991. "Antioxidant Properties of Natural Hydroquinones from the Marine Colonial Tunicate. *Aplidium californicum. Free Radical Biol. Med.* 11:63–68.

CRONLUND, A. L., and WOYCHIK, J. H. 1987. "Solubilization of Collagen in Restructured Beef with Collagenases and Amylase." *J. Food Sci.* 52:857–860.

DRANSFIELD, E., and ETHERINGTON, D. 1981. "Enzymes in the Tenderization of Meat." Pp. 177–178. *Enzymes and Food Processing*, edited by G. G. Birch, N. Blankebrough, and K. J. Parker. London: Applied Science Publishers.

DUNAJSKI, E., KOWALSKA, Z., ROZLUCKA, E., SIONDALSKA, A., MICZKA, M., and SIKORSKI, Z. E. 1987. "Preparation of Standardized Microbiological Media from Marine Proteinaceous Materials of Low Suitability as Food (in Polish)." Unpublished research report. Technical University of Gdansk.

EISEN, A. Z., HENDERSON, K. D., JEFFREY, J. J., and BRADSHAW, R. A. 1973. "A Colagenolytic Protease from the Hepatopancreas of the Fiddler Crab, *Uca pugilator.* Purification and Properties." *Biochemistry* 12:1814–1822.

EISEN, A. Z., and JEFFREY, J. J. 1969. "An Extractable Collagenase from Crustacean Hepatopancreas." *Biochim. Biophys. Acta* 191:517–526.

ERIKSSON, C. 1975. "Method for Controlling the Ripening Process of Herring." Canadian patent 969419.

FAULKNER, D. J. 1990. "Marine Natural Products." *Natural Product Reports* 7:269–309.

FEHMERLING, G. B. 1973. "Separation of Edible Tissues from Edible Flesh of Marine Creatures." U. S. patent 3,729,324.

FERRER, O. J., KOBURGER, J. A., OTWELL, W. S., GLEESON, R. A., SIMPSON, B. K., and MARSHALL, M. R. 1989. "Phenoloxidase from the Cuticle of Spiny Lobster (*Panulirus argus*): Mode of Action and Characterization." *J. Food Sci.* 54:63–68.

GERSHON, D. 1992. "Genetically Engineered Foods Get Green Light." *Nature* 357:352.

GILDBERG, A., and ALMAS, K. A. 1986. "Utilization of Fish Viscera." P. 383. In *Food Engineering and Process Applications*, edited by M. Le Maguer and P. Jelen. New York: Applied Science Publishers.

GODFREY, T., and REICHERT, J. 1983. *Industrial Enzymology. The Application of Enzymes in Industry*. New York: The Nature Press.

GRANT, G. A., SACCHETTINI, J. C., and WELGUS, H. G. 1983. "A Collagenolytic Serine Protease with Trypsin-like Specificity from the Fiddler Crab, *Uca pugilator*." *Biochemistry* 22:354–356.

GREEN, J. H., PASKELL, S. L., and GOLDMINTZ, D. 1977. "Fish Peptones for Microbial Media Developed from Red Hake and from Fishery By-product." *J. Food Protect.* 40:181–186.

HAARD, N. F. 1992. "A Review of Proteolytic Enzymes from Marine Organisms and Their Application in the Food Industry." *J. Aquatic Food Products Technol.* 1(1):17–35.

HAARD, N. F., FELTHAM, L. A. W., HELBIG, N., and SQUIRES, J. 1982. "Modification of Proteins with Proteolytic Enzymes from the Marine Environment. Pp. 223–244. In: *Modification of Proteins*, edited by R. L. Feeney and J. R. Whitaker. Advances in Chemistry Series 198. Washington, DC: American Chemical Society.

HAARD, N. F., KARIEL, N., HERZBERG, G., and FELTHAM, L. A. W. 1985. "Stabilization of Protein and Oil in Fish Silage for Use as a Ruminant Feed Supplement." *J. Sci. Food Agric.* 36:229–238.

HAARD, N. F., SHAMSUZZAMAN, K., BREWER, P., and ARUNCHALAM, K. 1983. "Enzymes from Marine Organisms as Rennet Substitutes." Pp. 237–241. In *Uses of Enzymes in Food Technology*, edited by P. Dupuy. Paris: Lavoisier.

HALLERMAN, E. M., KAPUSCINSKI, A. R., HACHETT, P. B., FARAS, A. J. and GUISE, K. S. 1990. Gene transfer in fish. In: *Advances in Fisheries Technology and Biotechnology for Increased Profitability*, eds. M. N. Voigt and J. R. Botta, p. 35–49, Lancaster, PA: Technomic Publishing Co.

HEMPL, E. 1983. "Taking a Short-cut from the Laboratory to Industrial-Scale Production." *Infofish Marketing Digest* 4:30.

HEW, C. L. 1989. "Transgenic Fish: Present Status and Future Directions." *Fish Physiol. Biochem.* 7(1–4):409–413.

HOUDEBINE, L. M. and CHOURROUT, D. 1991. "Transgenesis in Fish." *Experientia* 47:891–897.

IN, T. 1990. "Seafood Flavourants Produced by Enzymatic Hydrolysis." Pp. 425–436. In *Advances in Fisheries Technology and Biotechnology for Increased Profit-

ability, edited by M. N. Voigt and J. R. Botta. Lancaster, PA: Technomic Publishing Co.

INOUE, K., YAMASHITA, S., HATA, J., KABENO, S., ASADA, S., NAGAHISA, E., and FUJITA, T. 1990. "Electrovaporation as a New Technique for Producing Transgenic Fish." *Cell Differentiation and Deve.* 29:123–128.

IYER, K. M., GOPAKUMAR, K., SHENOY, A., JAMES, M. A., and NAIR, M. R. 1978. "Peptone from Threadfin Bream (*Nemipterus japonicus* Block): Preparation and Stability as Microbial Growth Media." *FAO, Indo-Pacific Fishery Com. Proc.* 18th session. Section III, Pp. 520–526, Symposium on Fish Utilization Technology and Marketing in the IPFC Region.

JACOBSEN, F., and LYKKE-RASMUSSEN, O. 1984. "Energy-Savings Through Enzymatic Treatment of Stickwater in the Fish Meal Industry." *Process Biochem.* 19(5):165.

JOAKIMSSON, K. G. 1984. Enzymatic deskinning of herring (*Clupea harengus*). Ph.D. dissertation, Institute of Fisheries, University of Tromso, Norway.

KINSELLA, J. E., GERMAN, J. B., and SHETTY, J. 1985. Uricase from fish liver: Isolation and some properties. *Comparative Biochem. Physiol.* 82B:621–624.

KNORR, D. 1991. "Recovery and Utilization of Chitin and Chitosan in Food Processing Waste Management." *Food Technol.* 44(1):114–122.

KNORR, D. and KLEIN, J. 1986. Production and conversion of chitosans with cultures of *Mucor rouxi* and *Phycomyces blakesleeanus.* Biotechnol. Lett. 8(10):691–694.

KOŁODZIEJSKA, I., PACANA, J., and SIKORSKI, Z. E. 1992. "The Effect of Squid Liver Extract on Proteins and on the Texture of Cooked Squid Mantle." *J. Food Biochem.* 16(3):141–150.

KRISTJANSSON, M. M. 1990. "Research Activities at the Marine Biotechnology Centre in Denmark." Pp. 543–546. In *Advances in Fisheries Technology and Biotechnology for Increased Profitability,* edited by M. N. Voigt and J. R. Botta. Lancaster, PA: Technomic Publishing Co.

LEE, Y. Z., SIMPSON, B. K., and HAARD, N. F. 1982. "Supplementation of Squid Fermentation with Proteolytic Enzymes." *J. Food Biochem.* 6:127–134.

LONG, A., and HAARD, N. F. 1988. "The Effect of Carotenoprotein Association on Pigmentation and Growth Rates of Rainbow Trout, *Salmo gairdneri.*" Pp. 99–101. *Proceedings of Aquaculture International Congress,* New Brunswick: Aquaculture Association of Canada.

MACLEAN, N., PENMAN, D., and ZHU, Z. 1987. "Introduction of Novel Genes into Fish." *Bio/Technology* 5:257–261.

MANU-TAWIAH, W., and HAARD, N. F. 1987. "Recovery of Carotenoprotein from the Exoskeleton of Snow Crab, *Chionecetes opilio. Can. Inst. Food Sci. Technol. J.* 20:31–33.

MCELVILY, A. J., IYENGAR, R., and OTWELL, W. S. 1991. "Sulfite Alternative Prevents Shrimp Melanosis." *Food Technol.* 45(9):80–86.

MOHR, V. 1980. "Enzyme Technology in the Meat and Fish Industries." *Process Biochem.* 15:18–21, 32.

Naczk, M., Synowiecki, J., and Sikorski, Z. E. 1981. "The Gross Chemical Composition of Antarctic Krill Shell Waste." Food Chem. 7:175–179.

NAKAJIMA, I., and SIBUYA, K. 1989. "Deodorizing Substances from Marine Algae." *Bio-Industry* 6:524.

NIP, W. K., MOY, J. H., and TZANG, Y. Y. 1985. "Effect of Purging on Quality Changes of Ice-Chilled Freshwater Prawn, *M. rosenbergii.*" *J. Food Technol.* 20:9–15.

NORLAND, R. E. 1990. "Fish Gelatin." Pp. 325–333. In *Advances in Fisheries Technology and Biotechnology for Increased Profitability*, edited by M. N. Voigt and J. R. Botta. Lancaster, PA: Technomic Publishing Co.

OCKERMAN, H. W., and HANSEN, C. L. 1988. "Seafood By-products." Pp. 279–308. In *Animal By-Product Processing*. New York: VCH Publishers.

OLSON, R. L., JOHANSEN, A., and MYRNES, B. 1990. "Recovery of Enzymes from Shrimp Waste." *Process Biochem.* 25:67–68.

OPOKU-GYAMFUA, A., SIMPSON, B. K., and SQUIRES, E. J. 1992. "Comparative Studies on the Polyphenol Oxidase Fraction from Lobster and Tyrosinase." *J. Agric. Food Chem.* 40:772–775.

OREJANO, F. M., and LISTON, J. 1981. "Agents of Proteolysis and Its Inhibition in Patis (Fish Sauce) Fermentation." *J. Food Sci.* 47:198–203.

PAN, B. S. 1990. "Recovery of Shrimp Waste Flavourant." Pp. 437–452. In Advances in Fisheries Technology and Biotechnology for Increased Profitability, edited by M. N. Voigt and J. R. Botta. Lancaster, PA: Technomic Publishing Co.

PLEIDERER, V. G., ZWILLING, R., and SONNEBORN, H.-H. 1967. "Eine Protease vom Molekulargewicht 11,000 und eine trypsinähnliche Fraktion aus *Astacus fluviatilis.*" *Z. Physiol. Chem.* 348:1319.

RAA, J. 1990. "Biotechnology in Aquaculture and the Fish Processing Industry: A Success Story in Norway." Pp. 509–524. In *Advances in Fisheries Technology and Biotechnology for Increased Profitability*, edited by M. N. Voigt and J. R. Botta. Lancaster, PA: Technomic Publishing Co.

RAA, J., and GILDBERG, A. 1982. "Fish Silage: A Review." *Crit. Rev. Food Sci. Nutr.* 16:383.

RAKSAKULTHAI, N., LEE, Y. Z., and HAARD, N. F. 1986. "Influence of Mincing and Fermentation Aids on Fish Sauce Prepared from Male, Inshore Capelin, *Mallotus villosus.*" *Can. Inst. Food Sci. Technol. J.* 19:28–33.

RAKSAKULTHAI, N., and HAARD, N. F. 1992. "Correlation between the concentration of Peptides and Amino Acids and the Flavour of Fish Sauce." *ASEAN J. Food Sci. Technol.*" 7:86–90.

RAO, S. S. V., SARASWATHI, C. R., and DWARAKANATH, C. T. 1978. "Studies on the Utilization of Fishery Wastes for the Production of Microbiological Media. *FAO, Indo-Pacific Fishery Com. Proc.* 18th Session. Section III, pp.

364–369, Symposium on Fish Utilization Technology and Marketing in the IPFC Region.

RAO, S. S. V., DWARAKANATH, C. T., and SARASWATHI, C. R. 1980. "Preparation and Microbiological Evaluation of Bactopeptone from Shrimp Waste." *J. Food Sci. Technol.* 17:133–135.

REBECA, B. D., PENA-VERA, M. T., and DIAZ-CASTANEDA, M. 1991. "Production of Fish Protein Hydrolysates with Bacterial Proteases: Yield and Nutritional Value." *J. Food Sci.* 56:309–314.

REECE, P. 1988. "Recovery of Proteases from Fish Waste." *Process Biochem.* 23:62–66.

REVAH-MOISEEV, S., and CARROAD, P. A. 1981. "Conversion of the Enzymatic Hydrolysate of Shellfish Waste Chitin to Single Cell Proteins." *Biotechnol. Bioeng.* 23:1067–1078.

RINEHART, K. L., KISHORE, V., BIBLE, K. C., SAKAI, R., SULLINS, D. W., and LI, K. M. 1988. "Didemnins and Tunichlorin–Novel Natural Products from the Marine Tunicate *Tridemnum solidum.*" *J. Nat. Prod.* 51:1.

RITSKES, T. M. 1971. "Artificial Ripening of Maatjes-Cured Herring with the Aid of Proteolytic Enzyme Preparations." *Fishery Bull.* 69:647–654.

ROBERTSON, A. 1991. "The Chemistry and Biochemistry of Black Tea Production—The Non-volatiles. Pp. 555–601. In *Tea. Cultivation to Consumption,* edited by K. C. Wilson and M. N. Clifford. New York: Chapman & Hall.

RUIZ-HARRARA, J. 1978. "The Distribution and Quantitative Importance of Chitin in fungi." In *Proceedings of the 1st International Conference on Chitin/Chitosan,* edited by R. A. A. Muzzarelli and E. R. Parsen. Cambridge, MA: M.I.T.

SCHEUER, P. J. 1989. "Marine Natural Products and Biomedicine." *Medicinal Res. Rev.* 9:535–545.

SETO, A. 1990. "Overview of Marine Biotechnology in Japan and Its Commercial Application for Aquaculture." Pp. 525–541. In *Advances in Fisheries Technology and Biotechnology for Increased Profitability,* edited by M. N. Voigt and J. R. Botta. Lancaster, PA: Technomic Publishing Co.

SIKORSKI, Z. E. and NACZK, M. 1981. "Modification of Technological Properties of Fish Protein Concentrates. *Crit. Rev. Food Sci. Nutr.* 14:202–230.

SIKORSKI, Z. E., and NACZK, M. 1982a. "Changes in Functional Properties of Fish Protein Preparations Induced by Hydrolysis." *Acta Alimentaria Polonica* 8:35–42.

SIKORSKI, Z. E., and NACZK, M. 1982a. "Manufacturing of Chitin Preparations from the Offals Generated during Krill Processing." Polish Patent 114,387.

SIMPSON, B. K., and HAARD, N. F. 1984. "Trypsin from Greenland Cod as a Food Processing Aid." *J. Applied Biochem.* 6:135–143.

SIMPSON, B. K., and HAARD, N. F. 1985. "Extraction of Carotenoprotein from Shrimp Processing Offal with the Aid of Enzymes." *J. Appl. Biochem.* 7:212–222.

SIMPSON, B. K., and HAARD, N. F. 1987a. "Cold-Adapted Enzymes from Fish."

Pp. 495–527. In *Food Biotechnology*, edited by D. Knorr. New York: Marcel Dekker.

SIMPSON, B. K., and HAARD, N. F. 1987b. "Trypsin and Trypsin-like Enzymes from the Stomachless Cunner. Kinetic and other physical properties." *J. Agric. Food Chem.* 35:652–656.

SIMPSON, B. K., MARSHALL, M. R., and OTWELL, W. S. 1987. "Phenoloxidases from Pink and White Shrimp: Kinetic and Other Properties. *J. Agric. Food Chem.* 35:918–921.

SIMPSON, B. K., MARSHALL, M. R., and OTWELL, W. S. 1988. "Phenoloxidase from Pink and White Shrimp (*Paneaus setiferus*). Purification and Properties." *J. Food Biochem.* 12:205–217.

SIMPSON, B. K., SMITH, J. P., and HAARD, N. F. 1991. "Marine Enzymes." Pp. 1645–1653. In *Encyclopedia of Food Science and Technology*. New York: Wiley.

SKORUPA, K., and SIKORSKI, Z. E. 1992. "Laboratory-Scale Preparation of Peptones for Microbiological Purposes from Inedible Parts of Cod Fillets (in Polish)." *Medycyna Weterynaryjna* 48:279–281.

STEFANSSON, G., JOHANNESSON, J., and MAGNUSDOTTIR, E. 1987. "Removal of intestines from scallop muscle with enzymes." Internal report. Iceland Fisheries Laboratory, Reykjavik, Iceland.

STEFANSSON, G., and STEINGRIMSDOTTIR, U. 1990. "Application of Enzymes for Fish Processing in Iceland—Present and Future Aspects." Pp. 237–250. In *Advances in Fisheries Technology and Biotechnology for Increased Profitability*, edited by M. N. Voigt and J. R. Botta. Lancaster, PA: Technomic Publishing Co.

STEINGRIMSDOTTIR, U., and STEFANSSON, H. 1988. "Enzymatic Removal of Swim Bladder Membrane." Internal report. Iceland Fisheries Laboratory, Reykjavik, Iceland.

STROM, T., and RAA, J. 1991. "From Basic Research to New Industries Within Marine Biotechnology: Successes and Failures in Norway." Pp. 63–71. In *Proceedings of Seminar on Advances in Fishery post-Harvest Technology in Southeast Asia*, edited by H. K. Kuang, K. Miwa, and M. B. Salim, Thailand.

SUGIHARA, T., YASHIMA, C., TAMURA, H., KAWASAKI, M., and SHIMIZU, S. 1973. "Process for Preparation of Ikuna (Salmon Egg)." U. S. patent 3,759,718.

SYNOWIECKI, J., SIKORSKI, Z. E., and NACZK, M. 1991. "The Activity of Immobilized Enzymes on Different Krill Chitin Preparations." Biotechnology and Bioengineering 23:2211–2215.

SYNOWIECKI, J., SIKORSKI, Z. E., PIOTRZKOWSKA, H., and NACZK, M. 1982. "Immobilization of Enzymes on Krill Chitin Activated by Formaldehyde." Biotechnology and Bioengineering 24:1871–1876.

TAVE, D. 1988. "Genetic Engineering." *Aquaculture Mag.* 13:63–65.

TRZESINSKI, P. 1955. "Tanning of Fish Skins (in Polish)." *Reports Sea Fish. Inst. Gdynia* 8:335–374.

TRZESINSKI, P. 1960. "Enzymatic Preparations from Fish Entrails to Skin Bating (in Polish)." *Reports Sea Fish. Inst. Gdynia* 10(B):111–124.

TSUGITA, T. 1990. "Chitin/Chitosan and Their Application." Pp. 287–298. In *Advances in Fisheries Technology and Biotechnology for Increased Profitability*, edited by M. N. Voigt and J. R. Botta. Lancaster, PA: Technomic Press.

VECHT-LIFSKITZ, S., ALMAS, K. A., and ZOMER, E. 1990. "Microbial Growth on Peptones from Fish Industrial Waste." *Microbiology* 10:183–186.

WEIST, J. L., and KAREL, M. 1992. "Development of a Fluorescence Sensor to Monitor Lipid Oxidation. 1. Fluorescence Spectra of Chitosan Powder and Polyamide Powder After Exposure to Volatile Lipid Oxidation Products." *J. Agric. Food Chem.* 40:1158–1162.

WHEATON, F. W., and LAWSON, T. B. 1985. *Processing Aquatic Food Products.* P. 445. New York: Wiley.

WRAY, T. 1988. "Fish Processing: New Uses for Enzymes." *Food Manufacture* 63(7):48.

14

CONCLUDING REMARKS

Bonnie Sun Pan and Zdzisław E. Sikorski

THE SCOPE

This book presents the results of current research on the properties of proteins and NPN of marine fish and invertebrates and the possibilities of utilization of this knowledge for improving the use of marine resources. The main areas where rapid further progress is likely to be expected due to utilization of research on proteins and NPN are

- more efficient direct use of fish and invertebrates for human food
- further improvement of the sensory quality of seafoods
- increased utilization of inedible parts of the catch for technical purposes
- avoidance of environmental pollution by fishery wastes

INCREASING THE USE OF SEAFOOD PROTEINS FOR DIRECT HUMAN CONSUMPTION

Only a fraction of the total amount of the edible parts of the annual landings of fish and marine invertebrates is used directly for human food. The utilization of edible animal protein resources for feeding slaughter animals must be regarded on a global scale as a waste because of the loss incurred in the conversion of the fish protein into pork or chicken.

The reasons for the actual inefficient utilization of marine resources for direct human consumption are related to the local economic and technical conditions, for example,

- too-low price for the less-valuable species
- lack of handling, processing, and storage capacity in the fishing industry to cope with seasonally large catches
- very stringent regulations of the local health authorities regarding qualification of the fish as fit for human consumption, for example, in cases of infestation with even harmless parasites
- lack of proper value-adding processing know-how and techniques, and of disseminating of existing knowledge which are necessary to manufacture high-quality seafood products from abundant but not very highly valued species.

The last mentioned problem can be solved by intensifying basic and technologically oriented research on the properties of fish proteins. Examples of successful implementation in the fishing industry of results of research on proteins are several processes of manufacturing various food products from Antarctic krill and of the use of many species of fish, including those of high fat content, for the production of surimi. However, there are still numerous projects in progress aimed at reducing the high waste of soluble proteins and NPN in conventional surimi manufacturing. Furthermore, better understanding of the characteristics of individual proteins of aquatic animals, as well as of the enzymatic and chemical procedures leading to selected modifications of the functional properties of these proteins and of their interactions with other food components, may provide the necessary know-how for the wider use of new generations of surimi or other seafood-protein preparations in the food industry.

A large increase in the world supply and of human use of fish protein will probably be achieved by further rapid growth of mariculture of the most valuable species of fish and shellfish. In the nearest future, the increase in the productivity in mariculture should result primarily by utilizing the effects of applied research in molecular biology. The outcome of such research, however, may also bring effects in form of modified functional properties of the proteins of aquatic animals.

Last but not least, there is a possibility of extending the use, as human food, of edible parts other than the muscle meat of aquatic organisms, in the form of various food products. In the Western countries, a limited number of fish species has been recognized as suitable for commercial exploitation and has been used primarily in a few traditional forms of preparation. Furthermore, generally the fillets and steaks of fish are regarded as suitable for human consumption, although, recently, minced flesh separated from frames after filleting is also used as well as flesh recovered from fish heads. On the other hand, according to Asian tradition, many other species and valuable parts of fish and marine invertebrates as dried, fermented, or cooked products have been consumed with great relish. Salmon milt in an unblemished sac and creamy white in color from wild coho or silver salmon has become a new product exported from North America to Taiwan. Salmon roe from the Yukon River chum was sold at U.S. $14.00/lb in the wholesale market of Tokyo. Salmon livers with liquid alder smoke and garlic in olive oil are canned in Oregon (USA). It is possible that many such products, prepared in controlled conditions, will find their way into the Western cuisine, just as has happened with several traditional soybean products from the East.

IMPACT OF PROTEINS ON THE QUALITY OF SEAFOODS

The quality of seafoods is affected by the nature of the raw materials, handling on board, preservation after catch, proper choice and application of the recipes, suitability of processing procedures, precision of the control of the processing parameters in the industrially used equipment, choice of packaging materials and of methods of their application, as well as the effectiveness of preventing undesirable changes during storage of the products.

There is a high demand for premium quality fish in many markets. The sale of tuna for sashimi, that is, top-quality fish, increased from 909,000 tons in 1975 to 1,321,000 tons in 1987 and the consumption of sashimi-tuna in the Tokyo area was 1.3–1.6 kg/capita in that year. The expensive tuna species, that is, bluefin, southern bluefin, and big eye tuna contain more pigment and lipid in the muscles than yellowfin and albacore. These quality attributes deteriorate easily due to oxidation occurring in postharvest handling or at the landing site. The price can be doubled if the quality is desirable for the sashimi market. This consumption demand has become the incentive for the tuna fisheries to use tuna longline fishing and to switch to ultralow-temperature freezing to preserve the red color, the integrity of muscle structure, and the water-holding capacity of the myofibrillar proteins. As the demand for fish of such high-quality grew, the capacity of ultralow-temperature freezing also increased several times during the recent decade.

The progress in fish processing results from the developments in the processing machinery and packaging materials and equipment, as well as from the effects of research on the behavior of the raw materials in various processing conditions. The components of fish, primarily responsible for the quality of fresh seafoods and of seafood products, are the lipids and nitrogenous compounds, that is, their native properties, changes due to storage and processing, and interactions with other muscle components. During the recent decade, especially, the impact of the lipid–protein interactions on the quality of seafoods has been stressed. The involvement of proteins in many processes that have a large effect on the quality of seafoods has been recognized. There are still, however, several problems awaiting further investigations, for example:

- the abnormal temperature effect on the dephosphorylation of ATP and on the onset and duration of rigor mortis in many species of fish
- the rate of autolytic and deteriorative changes in NPN and proteins in different species, as affected by biological and environmental factors
- the activity of muscle tissue enzymes in various species in fish and invertebrates, as influenced by the annual biological cycle and the environmental conditions
- the presence and role of different enzyme inhibitors in the muscles of fish and invertebrates
- the biochemical and chemical processes responsible for the discolorations in the meat of fresh, frozen, and canned fish, mainly tuna
- the changes of collagen in the connective tissues of fish and marine invertebrates due to starvation, the biological cycle, and after catch
- the role of individual muscle proteins and of their interactions in gelation of fish and squid meat
- the mechanisms of toughening of fish meat due to frozen storage and of the action of different inhibitors on the textural changes in frozen stored fish
- the interactions of fish proteins with the components of wood smoke during smoking.

Further progress in research in these and many other areas related to proteins of marine fish and invertebrates should be helpful in improving the quality of seafood products and in extending the use of functional seafood proteins in the food industry.

BETTER UTILIZATION OF SEAFOOD NITROGENOUS COMPOUNDS FOR TECHNICAL PURPOSES

Roughly one-third of the world's fish catch goes into international trade and amounts to U.S. $33,000 million. For many developing countries the net surplus from seafood exports over imports increased by nearly three times during the last decade. Therefore, fish is not only a source of protein but also a source of income. It is not only a natural resource but is more of a trade commodity and source of profit. Even fish offal, considered in one part of the world as waste, can be a source of value-added by-products to other markets.

There is an increasing awareness of the need to utilize more efficiently the proteins and NPN contained in various fish processing discards and effluents, to decrease the burden to the environment, and to produce different value-added technical products. In different world markets the price of frozen cod head may reach about 50% of that of frozen round cod, whereas 1 kg of frozen milt, roe, or liver may cost several times that of the whole fish.

Among the most promising projects regarding value-added non-food products from fish offal is the manufacturing of enzyme preparations, protein isolates of specific properties, including modified collagens, for cosmetic and medical applications, bacterial growth regulators for microbiological purposes, and a large variety of products made of chitin. Molluscan shell contains polyanionic proteins presumably capable of controlling the growth of carbonate biomineral. These proteins have a high affinity to bind calcium carbonate crystals and can be used as anti-scalants or dispersants required in water treatments (Wheeler, 1991). Marine organisms from deep-sea vents have provided thermostable DNA polymerase and endonucleases for recombinant DNA technology and luciferase for gene studies in mammalian cells. Ground fish inhabiting ice-laden seawater have antifreeze proteins in their serum (Fletcher et al., 1986; Davis and Hew, 1990). These proteins may be applied to produce transgenic fish with improved freeze–thaw stability (Fletcher et al., 1986; Fletcher et al., 1988) or simply to upgrade the quality of frozen surimi. Adhesive byssus proteins isolated from mussels are characterized by a serial array of peptide repeats that occur 80 or more times in the primary sequence (Filpula et al., 1990). Lysine and 3,4-dihydroxyphenyl-alanine are prominent in the peptide repeats (Waite et al., 1985). Lysyl- and tyrosyl-derived residues are in equimolar proportions. Glycine, serine or threonine, and proline or hydroxyproline are also abundant in these proteins (Rzepecki et al., 1991), resembling some of the features of collagen. If these proteins can be produced on a commercial scale by using biotechnology, they have the potential application as adhesives in underwater construction.

The trend to drastically reduce the escape of nitrogenous compounds into the environment with factory effluents can best be realized by rationalizing the technological processes in the fish industry, by more efficient use of all edible parts of the catch, by applying optimum parameters of wastefree converting the offals into various products for animal feeding, by processing other parts into technical products, and by decreasing the amount of water used in various stages of handling and processing fish. Finally, biotechnological processes can be applied to recover the proteinaceous components and NPN from waste waters.

REFERENCES

DAVIS, P. L., and HEW, C. L. 1990. "Biochemistry of Fish Antifreeze Proteins." *FASEB J.* 4:2460–2468.

FLETCHER, G. L., KAO, M. H., and FOURNEY, R. M. 1986. "Antifreeze Peptides Confer Freezing Resistance to Fish." *Can. J. Zool.* 64:1897–1901.

FLETCHER, G. L., SHEARS, M. A., KING, M. J., DAVIES, P. L., and HEW, C. L. 1988. "Evidence for Antifreeze Protein Gene Transfer in Atlantic Salmon (*Salmo salar*)." *Can. J. Fish. Aquat. Sci.* 45:352–357.

FILPULA, D. R., LEE, S. M., LINK, R. P., STRAUSBERG, S. L., and STRAUSBERG, R. L. 1990. "Structural and Functional Repetition in a Marine Mussel Adhesive Protein." *Biotech. Prog.* 6:171–177.

RZEPECKI, L. M., CHIN, S. S., and WAITE, J. H. 1991. "Molecular Diversity of Marine Glues: Polyphenoloc Proteins from Five Mussel Species." *Molec. Marine Bio. Biotech.* 1(1):78–88.

WAITE, J. H., HOUSELEY, T. J., and TANZER, M. L. 1985. "Peptide Repeats in Mussel Glue Protein: Theme and Variations." *Biochemistry* 24:5010–5014.

WHEELER, A. P. 1991. "Polyanionic Proteins: From Biomineral to Biodegradable Commercial Products." International Marine Biotechnology Conference, Abstract S–2.

INDEX

223